THE
WEATHER
BOOK

THE WEATHER BOOK

Project managed by: Dynamo Limited
Author: Steve Parker
Consultants:
Matthew Reay, Meteorologist;
Rachel Reay, Meteorologist
Publisher: Piers Pickard
Editorial Director: Joe Fullman
Editor: Christina Webb
Art Director: Andy Mansfield
Design and illustrations: Dynamo Limited
Print Production: Nigel Longuet

Published in October 2022 by Lonely Planet
Global Ltd
CRN: 554153
ISBN 978-1-83869-530-9

Printed in Malaysia
10 9 8 7 6 5 4 3 2 1

STAY IN TOUCH:
lonelyplanet.com/contact

Lonely Planet Office:
IRELAND
Digital Depot, Roe Lane (off Thomas St.),
Digital Hub, Dublin 8, D08 TCV4

Lonely planet KIDS

THE WEATHER BOOK

STEVE PARKER

CONTENTS

INTRODUCTION

Weather rules our lives in endless ways. Warm, sunny days send people to the beach, while windy ones can turn turbines to create clean, sustainable energy. But droughts mean failed crops and less food, and wildfires destroy everything in their path. Furthermore, due to human activities, the weather and climate are changing. This is a gigantic challenge facing everyone around the world.

SNOW, ICE, AND SPEED

Weather in cold, high places means plenty of snow and ice. Such conditions are challenging and fewer people live here. But millions visit for leisure and sports such as skiing, ski-jumping, snowboarding, sledding, snowmobile racing, and more. Every four years, the Winter Olympics celebrates cold-weather sports, such as downhill skiing.

WEATHER AS INSPIRATION

The power and beauty of spectacular weather has long inspired artists and other creative people. Since ancient times, looming clouds, angry storms, colorful rainbows, red skies at dawn and dusk, giant waves, eerie fog and mist, scary whirlwinds, and many other weather events have been drawn and painted. The 19th-century British artist J.M.W. Turner was particularly inspired by weather.

The Fighting Temeraire (1839) by J.M.W. Turner shows ships beneath a glowing sunset.

THE WEATHER IS CHANGING

Human activities, such as burning fossil fuels, are causing Earth to gradually get warmer. This is changing the climate—the patterns of weather over decades and centuries. Extreme weather events, such as storms, droughts, floods, and wildfires, will become more common. We can all help lessen this global problem.

WHAT TO WEAR TODAY?

Weather affects the clothes we wear. Thick coats, hats, scarves, and gloves are essential in places with cold winters. Summer ushers in shorts and T-shirts, often worn with a hat and sunscreen. In very hot, sunny places, such as Dubai, in the United Arab Emirates (above), people wear traditional cotton garments. These are long, shading much of the body, but also loose enough to allow cooling air to circulate.

Vehicles attempt to make their way through a flooded street in Dhaka, Bangladesh.

WEATHER FROM SPACE

Right now, high in space, there are weather satellites orbiting our planet. Known as meteorological satellites, they take thousands of pictures and make millions of measurements every day. Their information is used to compile weather forecasts to help us organize a sunny day out—or warn us that a fierce, deadly hurricane is on the way.

DANGEROUS WEATHER

In an average year, about 100,000 people die in accidents related to weather, such as floods and blizzards. That's about one person every five minutes.

EVERYDAY WEATHER

Sometimes days go past with not much change in the weather. It's cool and cloudy, or warm and sunny, and not much else happens. This can continue for many weeks. All places around the world have their own "everyday weather." It may well not be very exciting, but it is predictable.

At any place on our planet, more than a hundred different features affect the weather. These include the Sun's warmth, the times it rises and sets, how much water is in the air, the wind's strength, the shape of the land, and the kinds of plants living in a particular place. The location of a place on Earth is also vital—whether it is near the sea or inland, how close it is to the Equator or the North or South Pole, how high up it is… the list goes on. But most places have their regular weather patterns throughout the year.

READY TO PLOW
Regular winter snows mean that roads, railroads, airports, and streets have to be cleared as soon as possible. So as the cold season approaches, snowplows are made ready for urgent action. Problems could mean that people become stuck for a time on their travels… or worse.

RAINING AGAIN

Wet days can make some people irritated and depressed. London, in the United Kingdom (below), has 120 rainy days each year and New York City has 150. Cairo in Egypt has only 15, but Quibdo in Colombia has 20 times more, at 300. People living in these places know what to expect, day after day.

THE ATMOSPHERE

Weather happens in air. Our planet's blanket of air is called the atmosphere. It is thickest (most dense) and heaviest near the surface, which is where most weather occurs. The atmosphere gradually fades with height. By 62 miles (100 km) above the surface, the air has dwindled to virtually nothing. This is where outer space begins. Since there is no air in space, there is no weather either.

EXOSPHERE
62,000 miles (100,000 km)

HIGH-ORBIT SATELLITES

LAYERS OF THE ATMOSPHERE

The atmosphere is held around Earth by the planet's gravity. There are five layers: troposphere, stratosphere, mesosphere, thermosphere, and exosphere. The lowest is the troposphere, which holds three-quarters of all the air. Next up is the stratosphere, where passenger jet aircraft usually fly. The change between one layer and the next is called a "pause."

THERMOPAUSE: 310–620 miles (500–1,000 KM)

THERMOSPHERE

AURORAS

INTERNATIONAL SPACE STATION
250 miles (400 km)

LOW-ORBIT SATELLITES

KÁRMÁN LINE - generally accepted boundary of outer space – 62 miles (100 km)

MESOPAUSE: 50–62 miles (80–100 km)

MESOSPHERE

SPRITES (upper atmospheric lightning)

METEOROLOGICAL ROCKET

METEORS

STRATOPAUSE: 31–40 miles (50–65 km)

STRATOSPHERE

STRATOSPHERIC CLOUDS

WEATHER BALLOON

OZONE LAYER: 19–25 miles (5–40 km)

SPY PLANE

AIRLINER

TROPOPAUSE: 3.7–12.5 miles (6–20 km)

TROPOSPHERE

MT. EVEREST
5.5 miles (8.8 km)

RÜPPELL'S GRIFFON VULTURE
(highest-flying birds): 6.8 miles (11 km)

JET STREAM
4–9 miles (7–15 km)

BEAUTIFUL VIEW

From a jet aircraft cruising at about 7.5 miles (12 km) high, the clouds look beautiful and peaceful. But just outside the window, the temperature is -58°F (-50°C) and the air is five times less dense than at ground level.

COLDER THEN WARMER

Temperature decreases as you move up through the troposphere, but then increases in the stratosphere. This is due to the way the Sun's energy is absorbed by the stratosphere's ozone layer, made up of a form of oxygen gas called ozone.

WHAT'S IN AIR?

Air is a mixture of invisible gases with no taste or smell. Almost four-fifths of air is nitrogen. Nearly one-fifth is oxygen, needed for life. In much smaller amounts, but more important for weather, are carbon dioxide and the gas form of water known as water vapor. Carbon dioxide plays a major role in the greenhouse effect (see pages 156–159), while water vapor turns back into liquid water or ice to make clouds, mist, fog, rain, hail, and snow.

This diagram shows the proportions of gases in cold, dry air—that is, without water vapor. Warmer air is able to hold more water vapor. So in humid or misty air, the proportion of water vapor may be as high as 4 percent.

Nitrogen 78%

Oxygen 21%

Argon, other gases almost 1%

Carbon dioxide 0.04% (and rising)

SUN POWER

The Sun powers almost all weather on our planet. Its light, heat, and other rays are known as solar radiation or solar energy. When this reaches the atmosphere, some of the rays are taken in, or absorbed, by the air, including some of the heat. Other rays, such as light and the rest of the heat, travel all the way to the surface. It is the Sun's heat, called infrared radiation, that has most effect on our weather.

WHAT IS IN SUNSHINE?

The Sun is our nearby star, 93 million miles (150 million km) away. Its rays, or radiation, take about 8 minutes 20 seconds to reach Earth. There are many kinds of rays in solar radiation. Heat has the longest waves, while X-rays have the shortest.

HEAT (INFRARED)

LIGHT

ULTRAVIOLET

X-RAYS

HEAT (INFRARED)

Heat makes up one-half of solar radiation energy. Some heat is absorbed in the atmosphere, especially by the ozone layer. The rest reaches the surface.

LIGHT

Making up about two-fifths of solar radiation, almost all light passes through the atmosphere to the surface.

ULTRAVIOLET

Forming about one-twelfth of solar radiation, almost all of these rays are absorbed by the upper atmosphere.

X-RAYS

X-rays make up about one-hundredth of the total solar radiation. They are absorbed by the time they reach the mesosphere (see page 10).

SUNRISE BRINGS WARMTH

As the Sun rises in the morning, its heat quickly affects the surface. Moisture such as puddles and dew begin to turn into mist and water vapor, and are soon gone. When heat makes liquid water turn into water vapor, this is known as evaporation. Over a longer time, the Sun can dry up whole lakes, rivers, and even seas.

HOT AND COLD

The shape of the Earth has huge effects on how much of the Sun's rays reach a particular area. Around the middle of the planet, the Tropics, the rays shine straight down through the atmosphere onto a small area. Solar heat is powerful and concentrated, and so the area is hot. Near the poles, in the Arctic and Antarctic Circles, the rays hit at an angle and cover a larger area. Solar heat is more spread out, and the area is cooler.

SUN'S RAYS

In wintery polar lands, the Sun is low in the sky even at midday.

NORTH POLE

SUNLIGHT STRIKES AT AN ANGLE

SUNLIGHT STRIKES DIRECTLY

THE TROPICS

THE TROPICS

SUNLIGHT STRIKES AT AN ANGLE

SOUTH POLE

In the hot tropics the Sun is high overhead at midday.

UNWELCOME RAYS

The Sun's ultraviolet rays can cause great harm to living things, especially those not used to strong sunlight. This includes humans. In most places, bright sunshine means wearing hats, shirts, and sunscreen. Over-exposure to harmful rays has serious risks, and can cause skin growths and cancers.

Water evaporates into the air due to heat from the morning Sun.

WHY THE WIND BLOWS

The Sun's heat, or infrared rays, warms our planet unevenly. The heat is stronger around the Tropics than at the poles because its rays come almost straight down, rather than at an angle. Also, in all regions, the rays are soaked up by dark ground, which becomes warmer, but are reflected by light, bright surfaces like ice, snow, and clouds. In particular, water takes in heat more slowly than land. These differences affect the air above the surface and set it moving, as wind.

SHAPED BY WIND

Some places have constant strong winds that come mainly from one direction, especially along the coast or in uplands. Some plants can grow up shaped by the wind, leaning in one direction, such as these trees on the southern coast of South Island, New Zealand. Wind direction is named after where it blows from. So a westerly wind comes from the west and blows to the east.

HOW A SEA BREEZE HAPPENS

It's usually breezy at the coast. The Sun's heat warms the land faster than the sea. The warm ground heats the air above it. This warm air expands, becoming larger and lighter. So it rises, like the warmth from a fire or a home radiator. Cooler, heavier air from above the sea moves along to take its place, and this is wind. This process happens almost everywhere as different surfaces warm up by different amounts.

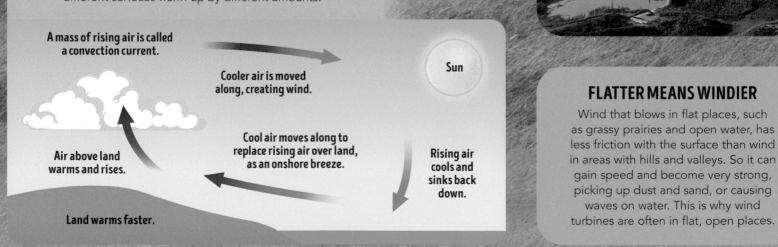

A mass of rising air is called a convection current.

Cooler air is moved along, creating wind.

Sun

Air above land warms and rises.

Cool air moves along to replace rising air over land, as an onshore breeze.

Rising air cools and sinks back down.

Land warms faster.

FLATTER MEANS WINDIER

Wind that blows in flat places, such as grassy prairies and open water, has less friction with the surface than wind in areas with hills and valleys. So it can gain speed and become very strong, picking up dust and sand, or causing waves on water. This is why wind turbines are often in flat, open places.

MEASURING WIND

Wind speed is measured in various ways:

➡ Miles or kilometers per hour (mph/km/h)

➡ Feet or meters per second (fps, m/s)

➡ Nautical miles, or knots, per hour (1 knot = 1.15 mph or 1.85 km/h)

➡ A useful measure is the Beaufort scale, invented by British Royal Navy officer Francis Beaufort in 1805.

THE BEAUFORT SCALE

BEAUFORT NUMBER	USUAL DESCRIPTION	WIND SPEED	LAND CONDITIONS	SEA CONDITIONS
0	CALM	Less than 1 mph, 1 km/h		Flat surface
1	LIGHT AIR	1–3 mph 1–5 km/h		Small ripples
2	LIGHT BREEZE	4–7 mph 6–11 km/h		Small, gentle waves
3	GENTLE BREEZE	8–12 mph 12–19 km/h		Small waves with foaming tops
4	MODERATE BREEZE	13–18 mph 20–28 km/h		Medium waves with foaming tops
5	FRESH BREEZE	19–24 mph 29–38 km/h		Waves with spray
6	STRONG BREEZE	25–31 mph 39–49 km/h		Large waves with big foaming crests
7	HIGH WIND, MODERATE GALE	32–38 mph 50–61 km/h		Spray whips across breaking waves
8	GALE	39–46 mph 62–74 km/h		Larger spray, waves roll together
9	STRONG OR SEVERE GALE	47–54 mph 75–88 km/h		Large waves crash together
10	STORM, WHOLE GALE	55–63 mph 89–102 km/h		Whole sea is waves, foam, and spray
11	VIOLENT STORM	64–72 mph 103–117 km/h		Waves higher than small boats
12	HURRICANE FORCE (SEE PAGE 108)	73 mph, 118 km/h or more		Boats tossed around and sunk

HIGHER MEANS WINDIER

Wind blowing along the ground rubs over the surface, and the friction slows it down. Higher up, there is less friction, so winds are faster. A wind at 3,300 ft. (1,000 m) high may be four times faster than the wind at ground level directly below. This is why wind turbines are often in high places.

AIR PRESSURE

Air is very light, but it is not weightless. Wherever there is atmosphere, its air presses in all directions on everything. This force is called air pressure or atmospheric pressure. Air pressure varies with air's temperature and how much water vapor it contains. Also, since air gets thinner and lighter with altitude, its air pressure lessens, too. The air in a typical school classroom at ground level weighs about 220 lb. (100 kg). If the classroom was at a height of 12 miles (20 km), its air would weigh about 10 times less.

CALM AND SETTLED WEATHER

High air pressure, known as an anticyclone, often means a period of settled weather, with light winds, few clouds, and little chance of rain.

UNSETTLED WEATHER

Deep low air pressure, known as a cyclone (see next page), brings changeable or unsettled weather with clouds, winds and rain, or snow, like these conditions on a highway in the USA.

HIGH AIR PRESSURE: ANTICYCLONE

Air flows away from areas of high pressure toward lower pressure, spinning clockwise (in the Northern Hemisphere) due to Earth's rotation. Air from higher up sinks to replace it, becoming warmer. The resulting "high," known as an anticyclone, often leads to settled weather, with few clouds and light winds.

As Earth spins and its surface moves around, this makes the air spin, too. It rotates clockwise in the Northern Hemisphere (the upper half of the world) and counterclockwise in the Southern Hemisphere.

Moving mass of air comes down from above and flows away

Sinking air

HIGH PRESSURE

LOW AIR PRESSURE: CYCLONE

Low pressure sucks in air from higher pressure around it, rises, and spins counterclockwise (in the Northern Hemisphere). As this air cools, it can hold less water vapor, which condenses into water as clouds, rain, or snow. The result is a "low," or cyclone, resulting in cloudy, wet, windy, changeable weather.

The Earth's rotation causes the air to spin counterclockwise in the Northern Hemisphere and clockwise in the Southern Hemisphere.

Rising air spreads out at height

Rising air

LOW PRESSURE

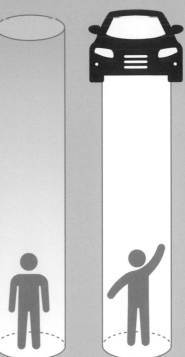

WEIGHT OF COLUMN OF AIR ABOVE US = **WEIGHT OF A SMALL CAR**

WHY WE ARE NOT CRUSHED

The column of air above a human body weighs about one ton (one tonne). Why do we not feel it pressing down on us? Air is gas, which is fluid and can flow. Air presses not just down but in all directions—sideways, upward, and at every other angle. These forces balance out, so we do not feel crushed by the air above us.

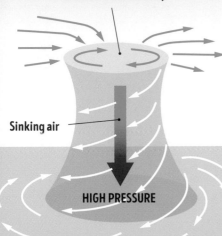

MEASURING AIR PRESSURE

Atmospheric (or air) pressure is measured in several ways:

→ **Atmosphere, atm**
The average or standard pressure at sea level is 1 atmosphere.

→ **Pascals, pa**
The standard unit of pressure in the metric system.

→ **Bars and millibars, mb**
Another unit of pressure used in weather forecasting.

→ **Pounds per square inch, psi**
A traditional unit of pressure in the imperial system.

→ **Millimeters of mercury, mm Hg**
This comes from air pushing on a thin column of the liquid metal mercury in an air pressure measuring device called a barometer.

1 atm = 101 pa = 1.013 bar = 1,013 mbar = 14.7 psi = 760 mm Hg

A traditional barometer, used to measure air pressure

WORLD OF WINDS

The Sun warms air and makes it move, which leads to wind developing. On a small scale, winds blow over a mountain or along city streets. On a huge scale, winds sweep across whole continents and vast seas. They are part of the wind patterns around the entire Earth, known as global air circulation. These winds do not blow in the same way all the time, every day, but they are the general direction of winds over years, decades, and centuries.

GLOBAL WIND PATTERNS

Between the Equator and the poles, the air is constantly rotating inside giant air masses called convection cells. Near the Equator, the Sun warms the air, which rises and cools—known as convection—and eventually spreads toward the two poles. But as it reaches the Tropics, this air sinks and travels back toward the Equator, creating an enormous cell, known as a Hadley cell. At the poles, the opposite happens. At the surface, the very cold air sinks and spreads toward the Equator, while warmer air high in the atmosphere moves poleward to replace it. This creates another enormous atmospheric cell. These two cells drive a third atmospheric cell in between. Together, these cells make up our global wind circulation. They dictate the usual pressure pattern and wind direction across the entire globe.

WINDS ACROSS THE OCEANS

Before engines, much long-distance travel and trade was by sailing ships. Sometimes at sea for months, sailors needed detailed knowledge of wind direction and strength. They called some winds trade winds because they helped the ships transport precious goods for business and trade. Due to the Earth's spinning, the trade winds to the north of the Equator are northeasterly, and southeasterly to the south of the Equator.

IN THE DOLDRUMS

On either side of the Equator, like a narrow belt around the middle of the world, is a region where the two sets of trade winds meet. It often has little or no wind. However, it does have clouds and rain. Its scientific name is the Inter-Tropical Convergence Zone (ITCZ), but it's usually known as the Doldrums, which means being miserable and going nowhere. Old-time sailing ships sometimes got stuck here for weeks, and that's how the sailors felt.

INTER-TROPICAL CONVERGENCE ZONE (ITCZ)

North Pole

POLAR CELLS

POLAR EASTERLIES

Subpolar low — 60°N

MIDLATITUDE CELLS

WESTERLIES

Subtropical high — 30°N

HADLEY CELLS

NORTHEAST TRADE WINDS

Equatorial low — EQUATOR

SOUTHEAST TRADE WINDS

HADLEY CELLS

Subtropical high — 30°S

WESTERLIES

Subpolar low — 60°S

MIDLATITUDE CELLS

POLAR EASTERLIES

POLAR CELLS

South Pole

WHEN NO WINDS BLOW

Many cities have polluting fumes from vehicles, factories, and buildings. These can form a foggy haze called smog (see page 31), as seen here in Mexico City. If this happens in a lowland area surrounded by hills (as here), there may be little wind to blow away the smog.

WAVES AND CURRENTS

"Dead calm" means no wind at all. Ponds, lakes, even seas can be flat and smooth, like mirrors, but the merest breath of wind makes a water's surface ripple slightly. Stronger winds cause bigger ripples. As the wind strength increases, the ripples become waves. As they grow, their tops, or crests, crash in a mass of spray and foam. Under the surface, the water is usually moving, too. These volumes of flowing water are known as currents.

WAVE FORCE

Water is heavy. A 3 ft. (1 m) cube of it weighs as much as a small car—that's about 1 ton (1 tonne). Blown into big waves by strong winds, water moves and crashes with tremendous force. It can gradually wear away the hardest rocks, batter strong harbor walls, as shown here in Porthleven, Cornwall, England, and even throw boats onto the land.

CURRENTS DUE TO GRAVITY

There are several kinds of water currents. One is caused by Earth's gravity pulling down on water, so that it flows downhill as a stream or river. The rushing water foams through rocky rapids and pours over waterfalls (right). Another type of current is due to the gravity of the Moon pulling on water as the Moon goes around, or orbits, the Earth. This makes seas and oceans rise and fall, known as tides, which creates tidal currents.

SUN-POWERED CURRENTS

As on land, the Sun heats oceans by different amounts in different places. For example, in the hot Tropics, water warmed by the Sun flows outward, cools, then sinks. Water from below rises up to replace it, is also warmed, flows outward, and so on. This sets up a circulation, or current, in that part of the ocean, similar to atmospheric cells in the air above (see page 18). Many other features affect this circulation, too (see page 22).

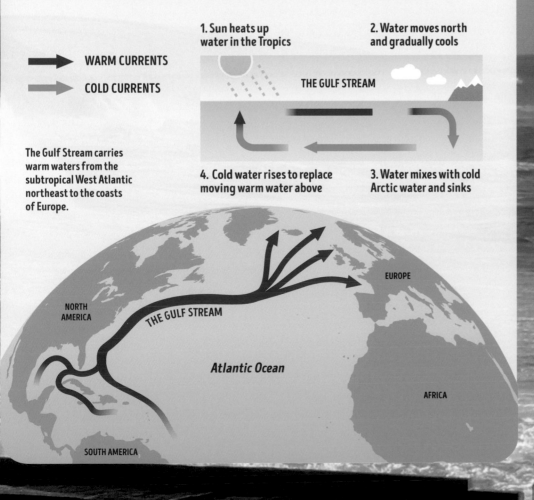

➡ **WARM CURRENTS**

➡ **COLD CURRENTS**

The Gulf Stream carries warm waters from the subtropical West Atlantic northeast to the coasts of Europe.

1. Sun heats up water in the Tropics

2. Water moves north and gradually cools

THE GULF STREAM

4. Cold water rises to replace moving warm water above

3. Water mixes with cold Arctic water and sinks

NORTH AMERICA

THE GULF STREAM

EUROPE

Atlantic Ocean

AFRICA

SOUTH AMERICA

DEADLY CURRENTS

Often it is not possible to see underwater currents from above, especially along the coast. When people travel to the coast to go swimming, or float on inflatables, there is a danger that currents, winds and waves may sweep them out to sea, requiring an emergency rescue. It is vital to look out for, and take note of, all warning signs about currents.

OCEAN CURRENTS

Seas and oceans are continually moving. Their waves and currents transport not only vast amounts of water, but also carry gigantic quantities of warmth or cold, too. Ocean currents move in all directions. They travel sideways, up and down, and at all angles in between. Like winds, ocean currents flow in regular, predictable patterns over years and centuries—and they have massive effects on the world's weather.

OCEAN CURRENTS AROUND THE WORLD

➡ WARM CURRENTS
➡ COLD CURRENTS

ATLANTIC OCEAN

ATLANTIC OCEAN

PACIFIC OCEAN

1.

2.

3.

4.

5.

SPEED AND DIRECTION

Where, how, and how fast ocean currents move depend on various factors:

1. The speed and direction of winds blowing across the surface of the water.
2. Changes in salt levels; for example, fresh water from a melting iceberg dilutes denser salty water around it.
3. The water's temperature (see pages 21, 23).
4. The Moon's gravity, which creates to-and-fro tidal currents.
5. The shape of coastline and seabed, which can change the direction of a current.

COOL CURRENTS

As well as the Sun's heat, freezing cold temperatures can also create currents. In polar regions, huge chunks of ice sheets split off into the ocean as icebergs (right). As they melt, their ice-cold water sinks fast. This draws in surface water from around, setting up currents that flow both sideways and vertically.

ARCTIC OCEAN

PACIFIC OCEAN

INDIAN OCEAN

SOUTHERN OCEAN

FINDING NEW HOMES

Long ago, ocean currents and global winds were an important factor in the way people spread around the world. In Southeast Asia and the Pacific, rafts and canoes carried people to islands where no humans had lived. There, they could set up new homes and communities.

A replica of an ancient Polynesian sailing canoe, used to settle the islands of the Pacific

CASE STUDY:

EL NIÑO AND LA NIÑA

The Pacific Ocean's winds, currents, and weather go through a regular change, or cycle. This cycle is caused by a complex combination of factors, including the way the Sun's heat warms different parts of the land and ocean, the changing winds and currents, and the shape of the coastlines. Every two or three years, a current called El Niño, "the boy," brings warmer waters to the west coast of South America. About two or three years later La Niña, "the girl," replaces it with colder water.

LOCATION

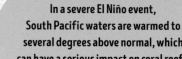

EL NIÑO, LA NIÑA, AND ENSO

El Niño and La Niña are part of a regular Pacific phenomenon, the El Niño Southern Oscillation (ENSO). El Niño's warm waters greatly affect South America, with more rain and even floods along southwest coasts. Farther north and east, from the Amazon to the Caribbean, it is drier and hotter, with possible droughts. La Niña years see more winter and spring rain in Southeast Asia, north and east Australia, while South America at this time has delayed and lighter rainfall.

DOUBLE THREATS

In a severe El Niño event, South Pacific waters are warmed to several degrees above normal, which can have a serious impact on coral reefs. Higher water temperatures damage coral, causing it to bleach and even die (see pages 162–163) —and many reefs are already suffering due to global warming (see page 158).

EL NIÑO YEAR

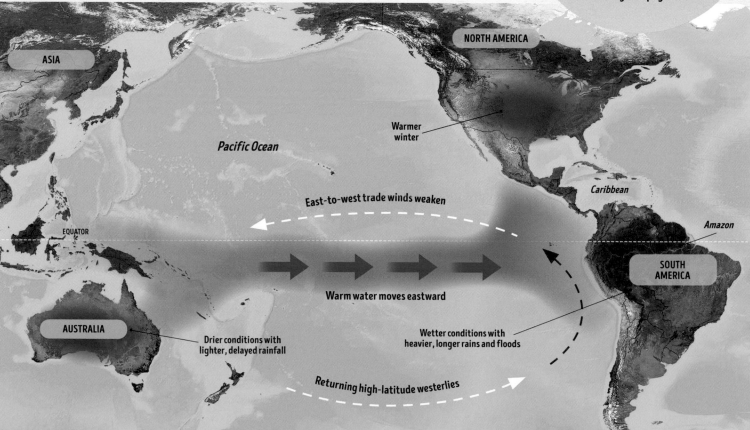

ASIA

NORTH AMERICA

Warmer winter

Pacific Ocean

Caribbean

Amazon

East-to-west trade winds weaken

EQUATOR

Warm water moves eastward

SOUTH AMERICA

AUSTRALIA

Drier conditions with lighter, delayed rainfall

Wetter conditions with heavier, longer rains and floods

Returning high-latitude westerlies

WEST EL NIÑO

In an El Niño period, the weather in Southeast Asia and eastern Australia is drier. Long periods of drought cause problems, such as parched crops and thirsty animals. Plant growers and farmers need to know when the "master weather-maker" El Niño is coming, so they can prepare.

A cracked, dried-out riverbed in Australia, caused by drought

EAST EL NIÑO

Fishing is a vital industry along the west coasts of North and South America. In an El Niño period, the main catch is sardines, which thrive in warmer waters. In La Niña years, there are more anchovies, since these fish prefer cooler waters. The effects of El Niño and La Niña are felt far away, too, even in Europe, Africa, and Asia.

Bunaken Island, Indonesia, which regularly endures the effects of El Niño events

LA NIÑA YEAR

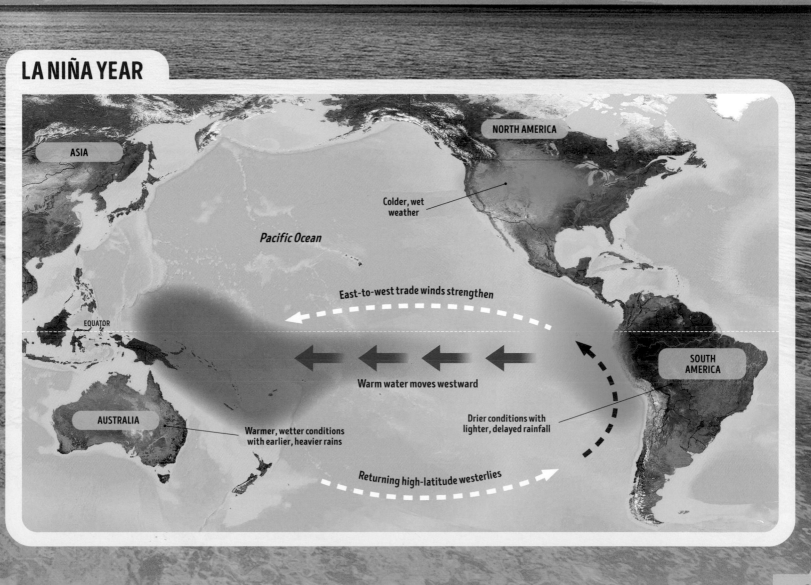

ASIA

NORTH AMERICA

Colder, wet weather

Pacific Ocean

East-to-west trade winds strengthen

EQUATOR

Warm water moves westward

SOUTH AMERICA

AUSTRALIA

Warmer, wetter conditions with earlier, heavier rains

Drier conditions with lighter, delayed rainfall

Returning high-latitude westerlies

WATER ON THE MOVE

Water is vital to life on Earth. It is also vital to our planet's weather. It may seem that newly made water falls from the sky as rain, then drains away and disappears forever. But new water is hardly ever made, or old water destroyed. It is the same water—endlessly moving around for thousands and even millions of years, traveling in a long and complicated journey known as the water cycle.

ROUND AND ROUND

The Sun warms liquid water in lakes and seas and on the land. This evaporates to become water vapor. It rises into the atmosphere, cools, and condenses—changes back into liquid water, usually as tiny floating droplets in clouds. The drops become bigger and heavier and fall as rain. Some rain runs along streams and rivers into lakes and seas, while some soaks into the ground and trickles along to emerge back onto the surface or into a lake or the sea. The cycle is never-ending.

Water falls as rain, ice, hail, or snow, known as precipitation.

Water vapor rises, cools, and condenses into cloud droplets.

Surface water, or runoff, flows into streams and rivers.

Plants give off water vapor in a process called transpiration.

Water soaks into soil and rocks as groundwater.

TRANSPIRE AND RESPIRE

Many living things give off water vapor into the atmosphere. Plant roots take up soil water and release it from their leaves as water vapor, a process called transpiration. Animal breath also contains water vapor, which condenses in cold air to form a steamy mist.

Water evaporates from seas, lakes, and rivers.

The Sun's warmth causes water to evaporate.

HOW LONG?

Scientists estimate that, on average, water spends this much time in different parts of the water cycle:

- ◊ **Ocean** – 5,000 years
- ◊ **Atmosphere** – 10 days
- ◊ **Rivers** – 2 weeks
- ◊ **Lakes** – 10 years
- ◊ **Living things** (such as plants and animals) – 1 week
- ◊ **Ice caps and glaciers** – 100–1,000 years
- ◊ **Deep in the ground** – 10,000–100,000 years

HIDDEN JOURNEY

Many small streams and brooks begin as water bubbling out of the ground, known as a spring. This water may have been in the ground, slowly flowing through soil and porous (spongy) rocks, for thousands of years.

HOT SPRINGS

Some water, like this hot water spring in Rotorua, New Zealand, doesn't need the Sun's heat to become water vapor. At hot springs, water in cracks and caves is heated by hot rocks deep underground. The water rises to the surface and gives off bubbles, vapors, spray, and steam that rise into the atmosphere.

27

CLOUDS

Endlessly moving, shape-shifting, coming, and going—clouds are among the most instantly recognizable features in the sky. They are also the best way to forecast the weather. Experts can recognize more than 50 different kinds of clouds, but most people can only tell the difference between around 10 main types. They vary in size, form, and color, and they float at different heights in the atmosphere. Small, puffy, cumulus clouds may exist for just a few minutes. Giant, looming hurricane clouds hang around for weeks.

CIRRUS

Thin, wispy, curly, feathery, made of tiny ice crystals. The highest clouds, at 19,500-plus ft. (6,000-plus m). **Indicate fair weather.**

CUMULONIMBUS

Massive towering clouds with puffy sides and an outward-spreading top, 1,150–39,000 ft. (350–12,000 m). They have a distinct shape with a flat top, caused by the clouds hitting the next layer of the atmosphere, the stratosphere. **Often bring rain and storms.**

NIMBOSTRATUS

Pale to dark gray sheet that spreads sideways, up, and down, 1,600–16,400 ft. (500–5,000 m). **Often brings drizzle or rain.**

CUMULUS

White or pale fluffy, puffy "cotton ball" clouds with flat bases. About 1,000–6,500 ft. (300–2,000 m). **Usually fine weather.**

CIRROSTRATUS

Widespread thin layer, whitish or milky, generally above 16,400 ft. (5,000 m).
Often means rain is on the way.

CIRROCUMULUS

Pale flakes, patches, tufts, and rows, sometimes with gray shading. Usually between 16,000 and 39,000 ft. (5,000 and 12,000 m).
Fair weather.

ALTOSTRATUS

Pale gray layers or sheets, perhaps faintly wavy. Medium height, about 6,500–19,600 ft. (2,000–6,000 m).
Possible light rain.

ALTOCUMULUS

Small, pale, puffy blobs and tufty lumps crowded together. Height about 8,200–18,000 ft. (2,500–5,500 m).
Rain unlikely.

STRATUS

Low, flat layers or sheets, white or gray, with few other features. Very low, usually below 1,500–2,000 ft. (450–600 m)
Drizzle or light rain.

STRATOCUMULUS

Many fluffy, puffy masses crowded or joined, gray underneath. Low, at 1,600–6,500 ft. (500–2,000 m).
Changeable weather.

MIST AND FOG

Clouds are usually found up in the sky, but sometimes they form in calm conditions at ground level, making it hard for us to see. When the visibility (how far we can see) is more than about 3,000 ft. (1,000 m), it is known as mist. When visibility is just a few hundred yards, it is called fog. Eventually, winds and the Sun's heat will make the conditions clear again.

TRAVEL PROBLEMS

Foggy weather causes all kinds of problems for travel. Cars and other vehicles must slow down so that drivers can stop within the distance they can see. Planes may be delayed unless they have the latest electronic aids for takeoff and landing in fog. Ships and boats move with caution, using powerful lights and loud horns and sirens, as they keep watch for flashing buoys and lighthouses. Trains also limit their speed.

TYPES OF FOG

Fog forms in many ways. Hill fog affects higher slopes, where water vapor in cooler air condenses while milder air in valleys below is clear. Evaporation fog appears when cold air moves over warm water, and the water vapor just above the surface condenses in the cold air. In San Francisco, moist, mild air from the Pacific Ocean is cooled as it meets colder, deeper water welling up along the coast. The moist air's water vapor condenses as coastal fog, and then onshore winds funnel it past the Golden Gate Bridge into San Francisco Bay (main picture). The effect is increased by warmed air from the cities around the bay rising and sucking in the cooler ocean air and fog. These persistent fogs are known locally as Karl.

EASILY LOST

Fog is often eerie and mysterious. It makes some people feel nervous, worried, and even lost, since they cannot recognize usual landmarks and views. It is also quiet, since it absorbs, or soaks up, sound. Often, people prefer to stay indoors until it clears.

SMOG

Smog is a mixture of smoke and fog. It is caused by air pollution from vehicle exhausts, power stations, factories, and other places that burn fuels. Some smog is made of tiny dust particles, around which water vapor has condensed as droplets. Photochemical smogs are made when sunlight changes gas chemicals known as nitrous oxides into solid particles. Antipollution laws are helping to make smog less common, but it continues to harm people, animals, and plants.

HAZY DAYS

Haze is not made up of water droplets but of very small, dry particles floating or drifting in the air. These particles come from many sources—natural dust from soil, fumes from industry and traffic, smoke from wildfires, and even bits of ash and fumes from volcanoes. Haze reduces visibility slightly, but usually we can still see for many miles.

WATER FROM THE SKY

The name for any form of water falling from Earth's atmosphere onto its surface is precipitation. This includes all kinds of rain and frozen water, such as hail, sleet, and snow. Rain clouds and rain are caused by warm, moist air (holding lots of water vapor) rising into the atmosphere, where it's cooler. The drop in temperature makes the water vapor condense into liquid water, which falls as rain.

Monsoon rain in the city of Hội An, Vietnam

SEASONAL RAINS

In many regions, rainfall is more common at certain times of year. In South Asia, this is usually June to September and is known as the summer monsoon season. It is due to moisture-laden winds blowing from the Indian Ocean in the southwest.

RAIN IS COMING

Rain clouds vary from less than 3 miles (5 km) across to hundreds of miles wide. When raindrops fall from clouds at a height of around 6,500 ft. (2,000 m), it takes about three minutes for them to splash onto the ground.

[○] Dramatic storm clouds move in across the desert as a thunderstorm dumps heavy rain in Arizona.

HOW RAIN HAPPENS

Rain clouds and rain are formed in several main ways, all as a result of warm, moist air rising into the cooler atmosphere above.

CONVECTIONAL RAINFALL

Land heated by the Sun causes air to rise (convection), which then cools, condenses, and forms cumulus or cumulonimbus clouds.

RELIEF RAINFALL

Warm, moist air blows against hills and mountains and is forced upward by their shapes and slopes.

FRONTAL RAIN

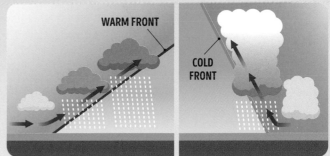

WARM FRONT

COLD FRONT

A cold front (see page 146) is the boundary between a warm air mass and a cold air mass. The cold air forces the warm air mass to rise, forming frontal clouds and widespread rain.

RAINDROP SHAPES

Raindrops are often imagined as teardrop or pear shapes, caused by the way they push through air, with the blunt end facing down and the narrow tapering tail at the top. In reality, they look quite different. Drops that are less than 0.04 in. (1 mm) across are round. As they grow, they become like burger buns, with a flat base and domed top. Larger ones are more like kidney beans or jelly beans, curving in slightly underneath.

<0.04 in. (<1mm)

0.07 in. (2mm)

0.12 in. (3mm)

DEW AT DAWN

As land cools during the night, the water vapor in the air just above it also cools, and condenses to form drops of dew. The drops can form on any object, including leaves, stones, and cars. They also cover and reveal items that are usually too small or thin to see, such as spiders' webs.

DROPLETS AND DROPS

An average tiny rain cloud droplet starts off at about 0.0007 in. (0.02 mm) wide. That's around one-quarter of the width of a human hair! As it grows (about 10 times larger), it becomes almost too heavy to float. By the time it has increased by another 10 times to around 0.0 in. (2 mm) wide, it is a raindrop and starts to fall. The largest raindrops are 0.3 in. (7–8 mm) across.

KAUA'I: MINI-RAIN SHADOW

As the low Sun shines on hills and mountains, it casts dark shadows on the region behind. Rain is similar. On the windward (upwind) sides of mountains, moist air blows up the slopes, cools into clouds, and drops rain. When the air reaches the mountain's other side, downwind (leeward), it holds much less moisture, with few clouds and little rain. This dry region is called a rain shadow. Some rain shadows are huge deserts, but there are also rain shadows in miniature, as on the northernmost Hawaiian island, Kaua'i.

LOCATION

Hawaii

📷 Sunrise on the southwest coast of Kaua'i in the rain shadow

DRY IN THE SHADOW

After the northeast winds pass over Kaua'i's central peaks of Mount Wai'ale'ale, they have much less moisture. The rain shadow affects the southwest of the island, around towns such as Kekaha and Waimea. Here, there is barely 20 in. (500 mm) of rain in a year—20 times less than on the mountain, only 20 miles (30 km) away.

MOUNTAIN OF RAIN

Kaua'i's high volcanic center, Mount Wai'ale'ale, has peaks rising to more than 5,000 ft. (1,500 m). It is one of the wettest places on the planet. Moisture-laden winds blow in from the Pacific Ocean onto its northeast slopes, bringing about 400 in. (10,000 mm) of rain each year.

Looking toward the cloud-covered mountains on Kaua'i's north coast

HAWAIIAN WINDS AND CLOUDS

Hawaii is in the middle of the Pacific Ocean. The main, or prevailing, northeast trade winds blow toward the southwest, as shown in the picture below.

HAWAIIAN ISLANDS

Kaua'i

Northeast trade winds

Rain shadow

KAUA'I

HAIL, FROST, AND SNOW

Few types of weather are as pleasing—or as disruptive—as snow. It can make the scenery beautiful and be great fun for sports and playing in. But snow can also bring travel and daily life to a standstill and leave people stranded in freezing, dangerous places. Regions in the far north and south of our planet, and also high up mountains, have many snowy days each year. Most places in the Tropics, around the middle of the planet, have never seen snow.

SNOWY SCENES

Fresh snow looks amazing and can be very useful. People love to roll in it, throw snowballs, build snow-people, and use sleds to ride on it. Some northern cities have snow almost guaranteed at a particular time of year and hold snow festivals, where wonderful snow and ice sculptures, such as gigantic life-sized castles and temples, are built. There's even an ice hotel built in Sweden every year.

[◎] The Ice and Snow Sculpture Festival in Harbin, Heilongjiang, China

HOW SNOW FORMS

Snow begins in a similar way to rain. Warm, moist air rises into the atmosphere where the temperature is below freezing. Its water vapor cools, but it does not condense into liquid water. Instead, it turns into ice crystals—a process known as deposition. The crystals grow until they become snowflakes, heavy enough to drift down to the surface.

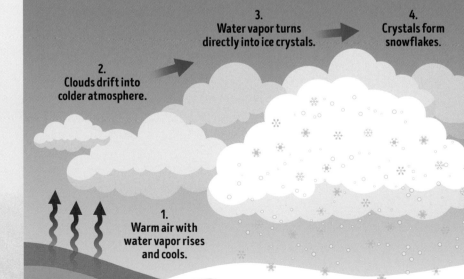

2.
Clouds drift into colder atmosphere.

3.
Water vapor turns directly into ice crystals.

4.
Crystals form snowflakes.

5.
Temperature is below freezing at all levels down to ground.

1.
Warm air with water vapor rises and cools.

EVERY ONE DIFFERENT

All snowflakes have six sides. This is due to the way that their microscopic building blocks, ice crystals, fit onto each other to grow. Yet it's thought that every snowflake has a unique shape. Most flakes are between 0.02 and 0.2 in. (0.5 and 5 mm) across.

FROST

Frost is made when water vapor in the air touches a very cold surface and the vapor turns into ice crystals. There are three kinds of frost: ground frost, which forms on the ground, objects, or trees, whose surfaces have a temperature below the freezing point of water; air frost, which occurs when the air temperature falls to or below the freezing point of water; and hoar frost (right), made from frozen dewdrops.

STONES FROM THE SKY

A hailstone begins as a water droplet in a storm cloud that moves into freezing atmosphere. The droplet becomes solid ice and may start to fall, but rising air below the storm cloud pushes it up again. More water attaches to the hailstone and freezes, making it bigger and heavier. This may happen several times before the hailstone has enough weight to fall. Large hailstones can damage vehicles, roofs, windows, and plants.

MORE KINDS OF COLD PRECIPITATION

FREEZING RAIN (left)
Rain that falls as liquid, but suddenly freezes in a very cold layer of air, just above or at the ground.

SLEET
Sleet often begins as snowflakes in a cloud. They fall and melt into raindrops as they pass through a layer of warmer air below, then freeze into lumps again in colder air near the ground.

SLUSH
Snow that is in the process of melting into soft, slippery lumps and pools.

WEATHER, SEASONS, AND CLIMATES

Weather happens hour by hour, day by day, week by week. A pattern of weather over a much longer time period (years and centuries) is known as a climate. Many regions also have annual seasons when the weather is cooler or warmer, wetter or drier, for parts of the year.

Sometimes, the weather changes in minutes, even seconds. It is bright with sunshine one moment, then quickly it becomes dull and cloudy, and starts to rain. In some places people say, "If you don't like the weather here, wait five minutes." Seasons are yearly. They are predictable, expected changes in weather throughout the months, like spring, summer, fall, and winter. Climate happens much more slowly. It is typical, or average, weather over very long periods of time, even thousands of years. Climates change very slowly, too—at least, they did long ago. Today, human activities are causing an unnatural and very rapid change in the world's climate (see page 160).

ICE AGES

Climates have changed many times during Earth's extensive history. The last great ice age began 110,000 years ago and lasted until around 11,000 years ago, when the planet began to warm up again. During this cold period, ice sheets and snow covered many northern lands, where woolly mammoths (left), woolly rhinos, giant cave bears, and other incredible, but now-extinct, creatures roamed.

LOCATION, LOCATION

Climates are usually colder in Earth's far north and south, and warmer at the Equator. However, land height also has a great effect on temperature, since the atmosphere is colder higher up. In Africa, Mount Kilimanjaro is just 200 miles (320 km) from the Equator. Temperatures on the surrounding savanna grasslands may exceed 86°F (30°C). Yet the mountain's peak, 19,350 ft. (5,895 m) above, has snow throughout the year.

ANCIENT RELATIVES

The African elephants that graze on the grasslands at the foot of Mount Kilimanjaro are distant relatives of woolly mammoths. They are much less hairy than their ancient ancestors because they live in a hotter climate. Mammoths died out partly because their frozen world vanished.

EARTH IN SPACE

The yearly seasons and their weather are due to how the Earth orbits the Sun. In particular, they are caused by the tilt, or leaning, of the Earth as it spins around once each day. At certain times of year, regions receive more of the Sun's light and heat, making warm or hot summers. At other times of the year, they receive less heat and light, making cool or cold winters. Around the middle of the planet, the Tropics, temperatures are much the same all year round. Seasons here are determined by how much rain falls.

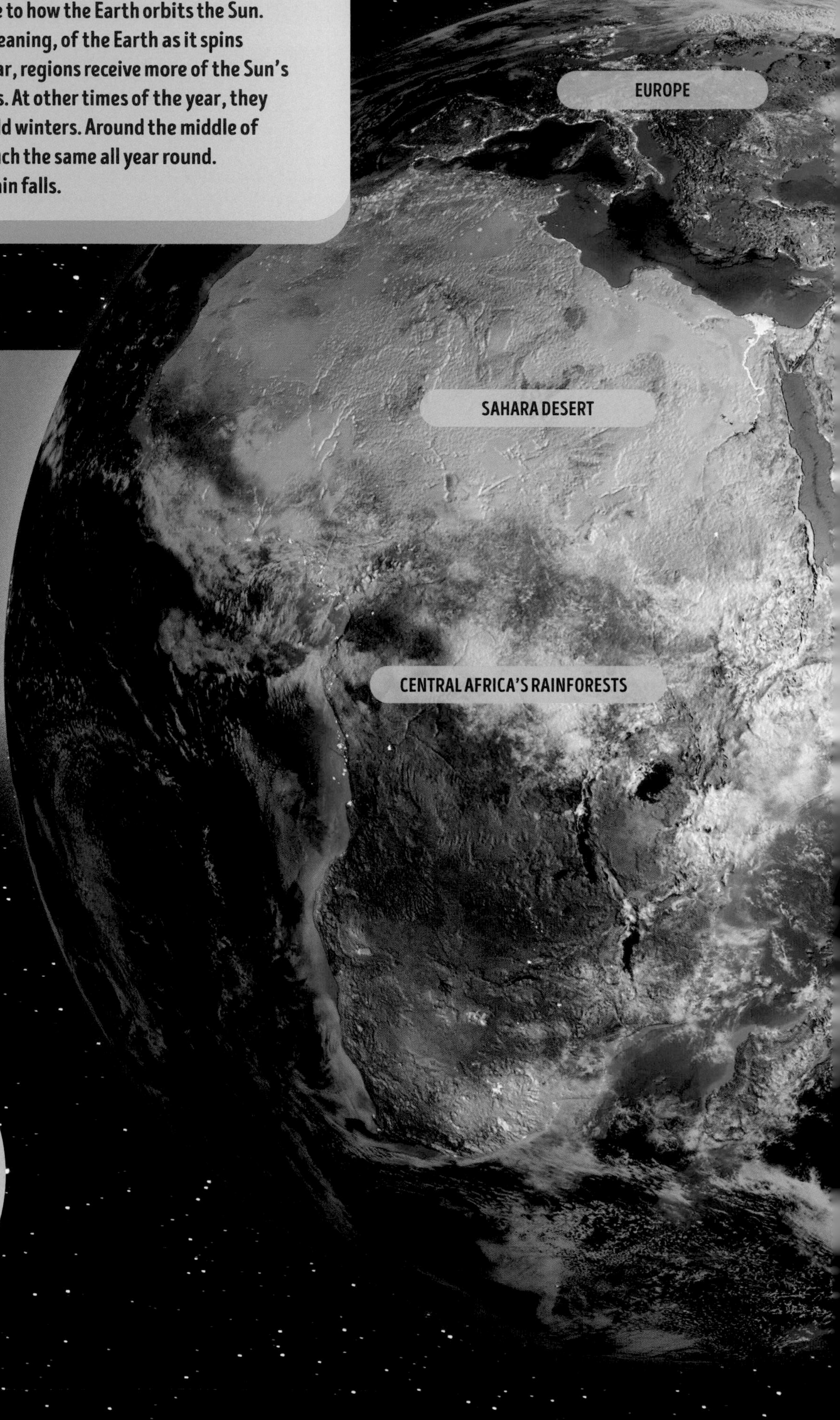

EUROPE

EARTH FROM SPACE

Images of Earth from space show the effects of seasons and climate. In this view, the deserts of Arabia and the Sahara in Africa have few clouds and a hot, dry climate all year round. South of the Sahara, you can see the clouds and greenery of Central Africa's tropical rainforests. To the north, you can see Europe, where most places have a moderate, or temperate, climate, with warm summers and cool winters.

SAHARA DESERT

CENTRAL AFRICA'S RAINFORESTS

FARTHER AND NEARER

Earth's orbit around the Sun is not a circle. It is shaped like an oval, or ellipse. The planet is closest to the Sun in early January, around 91 million miles (146 million km) away, and farthest away in early July, around 94.5 million miles (152 million km). These distances are not enough to affect the seasons.

EARTH'S YEARLY JOURNEY

As our planet orbits the Sun, it spins on its axis, an imaginary line between the North and South Poles. This axis is not upright, or vertical. Rather, the planet tilts, or leans, by 23.5°. For part of Earth's orbit, the Northern Hemisphere faces the Sun more directly. The Sun's rays come almost straight down, so the surface receives more heat and light, which is summer (and winter in the Southern Hemisphere). As the orbit continues, the Northern Hemisphere tilts away from the Sun. The heat and light come down at a shallow angle, more spread out and weaker, which is winter (but summer in the Southern Hemisphere).

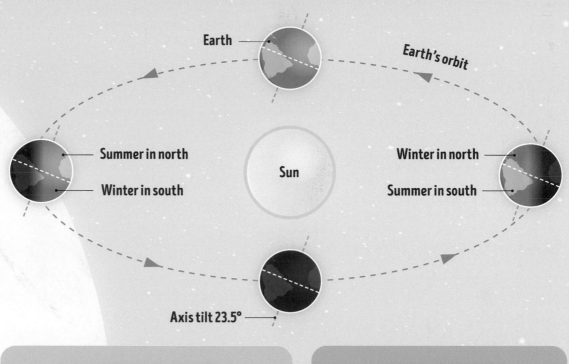

Earth

Earth's orbit

Summer in north

Winter in south

Sun

Winter in north

Summer in south

Axis tilt 23.5°

ARABIAN DESERTS

WINTER AND SUMMER

The winter solstice is when daylight is shortest and darkness is longest in a 24-hour day. In the Northern Hemisphere, this is around December 21–22. It is when the season of winter begins. Six months later, with daylight longest around June 21–22, the summer solstice marks the start of summer. The reverse happens in the Southern Hemisphere.

SPRING AND FALL

An equinox is when daylight and darkness are the same length in a 24-hour day. The vernal, or spring, equinox, March 20–21, is the start of spring in the Northern Hemisphere. The fall, or autumn, equinox, about September 22–23, signifies fall. Again, it is the reverse in the Southern Hemisphere.

SUN NEVER SETS

In the Antarctic, around the South Pole, it is midsummer in December. During this period, the Sun doesn't set for months. Even at midnight, the Sun dips low in the sky, then begins to rise again. The region has sunlight for 24 hours each day. In midwinter, June, the opposite happens and it is dark for several weeks.

CLIMATE ZONES

Climates around the world depend on many factors. Due to the planet's rounded shape, it is colder in the far north and south, and warmer near the Equator. Altitude (height above sea level) is also important, as the temperature falls about 3.5°F for every 1,000 ft. (6.5°C per 1,000 m). Distance from the sea has a great effect, too, because land warms up and cools down faster than water. This means that climates toward the center of the great landmasses, the continents, have more extremes of temperature and rainfall, while coastal climates are more moderate, or temperate.

MAIN CLIMATE ZONES

There are different systems for naming and mapping climate zones. Some systems have only a few zones; others have many. The detailed version of one of the main systems, the Köppen Climate Classification, has more than 30 types of climates for South America alone. This map uses a much simplified version of the Köppen system based on latitudes.

KEY

- POLAR AND SUBPOLAR ZONE
- TEMPERATE ZONE
- SUBTROPICAL ZONE
- TROPICAL ZONE

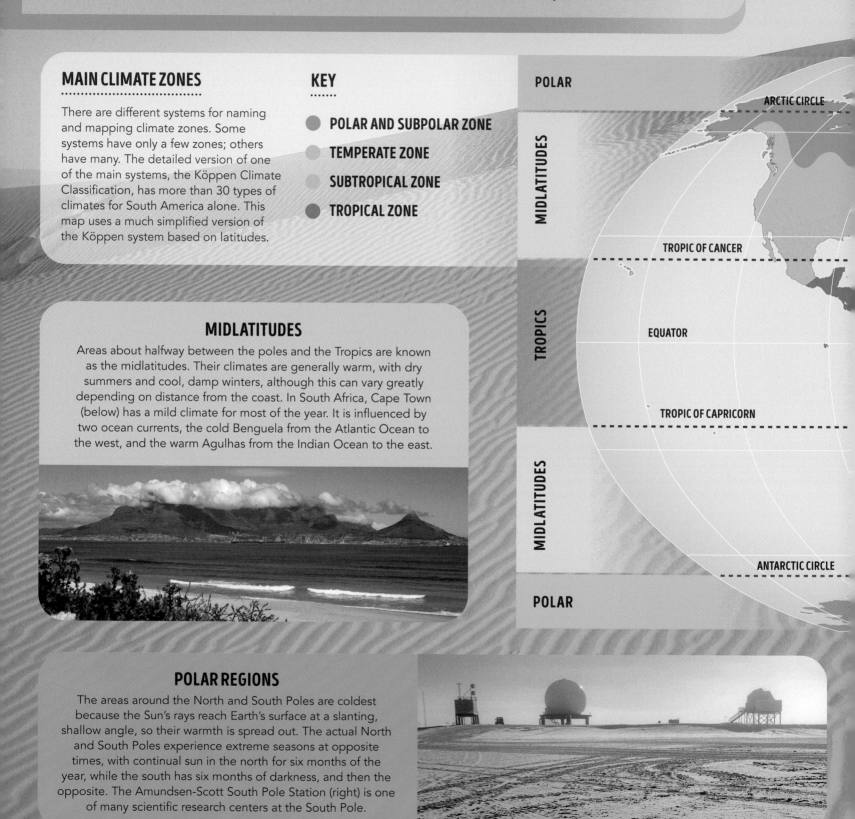

POLAR

ARCTIC CIRCLE

MIDLATITUDES

TROPIC OF CANCER

TROPICS

EQUATOR

TROPIC OF CAPRICORN

MIDLATITUDES

ANTARCTIC CIRCLE

POLAR

MIDLATITUDES

Areas about halfway between the poles and the Tropics are known as the midlatitudes. Their climates are generally warm, with dry summers and cool, damp winters, although this can vary greatly depending on distance from the coast. In South Africa, Cape Town (below) has a mild climate for most of the year. It is influenced by two ocean currents, the cold Benguela from the Atlantic Ocean to the west, and the warm Agulhas from the Indian Ocean to the east.

POLAR REGIONS

The areas around the North and South Poles are coldest because the Sun's rays reach Earth's surface at a slanting, shallow angle, so their warmth is spread out. The actual North and South Poles experience extreme seasons at opposite times, with continual sun in the north for six months of the year, while the south has six months of darkness, and then the opposite. The Amundsen-Scott South Pole Station (right) is one of many scientific research centers at the South Pole.

COASTAL AND CONTINENTAL CLIMATES

Lisbon, capital of Portugal, in Europe, has a mild coastal climate. The midsummer average is 75°F (24°C), falling to 52°F (11°C) in winter—a difference of 23°F (13°C). About the same distance north of the Equator is Bishkek, capital of Kyrgyzstan, in Central Asia. Its continental climate reaches 90°F (32°C) in summer and falls to 34°F (1°C) in winter, a much more extreme range of 56°F (31°C).

December: Lisbon, Portugal

December: Bishkek, Kyrgyzstan

Lisbon, Portugal

Bishkek, Kyrgyzstan

Borneo

Cape Town, South Africa

Amundsen-Scott South Pole Station

THE TROPICS

The Tropic of Cancer is an imaginary line north of the Equator, where the Sun is directly overhead for at least one day each year. The Tropic of Capricorn is similar, but south of the Equator. The area between these two lines is known as the Tropics. Tropical climates are mostly warm all year, often with a rainy season. The island of Borneo (left) in Southeast Asia is on the Equator itself. It is hot and humid almost all year, with a daily temperature of between 81°F and 86°F (27°C and 30°C). Its rainforests are home to some of Earth's greatest diversity of life.

POLAR CLIMATES

The North Pole and South Pole both have very cold climates, but they are very different. The North Pole is in an ocean, the Arctic Ocean, surrounded by land. The South Pole is on land, Antarctica, surrounded by the Southern Ocean. The waters of the Arctic Ocean hold what little warmth they receive for longer, compared to Antarctica's ice-covered surface. Antarctica is also at a much higher altitude, and so even colder than the Arctic's sea level. On average, the Arctic region around the North Pole is 36°F (20°C) warmer than the Antarctic region around the South Pole.

AT THE NORTH POLE

At the North Pole there is floating, drifting, shifting ice, usually about 10–13 ft. (3–4 m) thick. The seabed below is more than 13,000 ft. (4,000 m) down. The actual location of the North Pole cannot be marked on the ice as it is always moving. Explorers, scientists, and even tourists visit by sleds or aircraft, but there are no permanent structures.

A pointer temporarily marks the position of the North Pole on July 15, 2016.

The Northern Lights seen from Iceland

NORTH POLAR ANIMALS

Bordering the Arctic Ocean is the world's largest island, Greenland. Here, there are polar bears, Arctic foxes and hares, and caribou. The largest land animals are musk oxen (above). They have long, thick fur to keep out the intense cold, and dine on grasses, twigs, bushes, and any other plants they can find. Even in midsummer, July, Greenland's temperatures reach only about 50°F (10°C).

SHIMMERING CURTAINS

At the North and South Poles, the Earth's magnetic field affects tiny particles sent out by the Sun, known as the solar wind. The wind's electron and proton particles interact with the magnetism around the poles to give off light as vast, waving, shimmering, glowing curtains, about 50–500 miles (80–800 km) above the surface. They are known as the Northern Lights, Aurora Borealis, and the Southern Lights, Aurora Australis.

THE TUNDRA

Tundra is the name of the open, treeless lands in polar regions, covered by snow and ice for part of the year. Underneath the thin surface is permafrost, a layer of permanently frozen soil. Only small, low bushes, grasses, sedges, and similar plants can grow here. Summer lasts just a few weeks, when the landscape becomes covered with flowers before the snow returns.

Summer comes to the tundra in Chukotka, Siberia, in the far east of Russia.

THE HIGH ANTARCTIC

The highest place in Antarctica is the peak of Mount Vinson (below), 16,050 ft. (4,892 m) above sea level. It is about 750 miles (1,200 km) from the South Pole, and so windy that the scarce snowfall is often blown away, leaving the top bare and rocky. The combined effects of its location, altitude, and wind mean that the temperature here is 58°F (32°C) lower than at sea level on the nearest coast. A cold day on Vinson's peak can easily be -58°F (-50°C).

CASE STUDY:

ANTARCTICA'S FROZEN WORLD

Over millions of years, the world's continents have drifted around the planet. About 450 million years ago, Antarctica was at the Equator. By 200 million years ago, it had drifted south to the midlatitudes, as shown by fossils of plants and even dinosaurs in its rocks. Continuing south, it became increasingly colder and is now mostly frozen. Officially, Antarctica has a desert climate, with less than 10 in. (250 mm) of precipitation each year, almost all as snow and ice. Covering the land to an average depth of 7,200 ft. (2,200 m), the ice represents around two-thirds of all the world's fresh water.

LOCATION

SOUTH POLAR ANIMALS

Apart from humans, the largest animals on the Antarctic mainland are birds—emperor penguins, caring for their eggs and chicks. But after the breeding season, they return to the sea. The waters surrounding Antarctica are rich in food, especially small, shrimp-like krill, which are a vital part of the food chain for penguins, fish, seals, and whales.

SHEETS AND BERGS

As snow falls on Antarctica, it squashes the huge ice sheets, which spread outward. At the edge of the land, the ice sheets float out onto the surrounding ocean. Chunks break off, or "calve," into massive icebergs, as seen here in Andord Bay, Graham Land, Antarctica. Floating with the currents, "bergs" may last 10 years or more, providing homes for seals, penguins, and seabirds before they melt in warmer seas.

CLIMATE CLUES

Climate scientists drill into the thick ice of Antarctica to pull out rod-shaped ice cores, some dating back as far as 800,000 years. These ancient samples contain tiny bubbles of preserved air as well as minute traces of minerals. Analyzing these provides information about climates in ancient times (see page 152).

TOURIST HOTSPOT

Antarctic tourism is a growing business. One of the more easily reached areas is the Antarctic Peninsula, a narrow north-pointing strip of land within 625 miles (1,000 km) of the tip of South America. People take short trips to view seals, whales, and penguins from RIBs (rigid-hull inflatable boats). However, the tourist season lasts only a few months before storms, snow, ice, and darkness return.

THE JET STREAMS

The weather and climate of an area are shaped by many features, including its latitude (its location between the Equator and the poles), its height above sea level, and its distance from the sea. The jet streams can also have major effects. These are a group of very high, very fast, narrow bands of wind that blow around our planet from west to east, at heights of about 5–10 miles (8–16 km). Jet streams greatly influence both short-term weather patterns and long-term climates, especially in midlatitude, temperate areas.

DRAGGING THE ATMOSPHERE

The Earth rotates once every 24 hours. Its ball shape means different locations on the surface move at different speeds. The Equator is fastest and the poles are slowest. These differences mean the atmosphere is dragged along faster near the Equator compared to farther away. This distorts, or twists, the global circulation (see page 18), helping to create the jet streams.

STREAMING AROUND EARTH

There are four main jet streams, each a few hundred miles wide. The north polar jet stream blows between northern polar and temperate regions at latitudes between 50°N and 60°N. The north subtropical jet stream occurs between temperate and tropical areas, at latitudes of 30–40°N. The south polar and south subtropical jet streams are similar, but occur between the Equator and the South Pole. Polar jets are usually 5–8 miles (8–12 km) high and very fast—perhaps more than 190 mph (300 km/h). Subtropical jets are higher, 6–10 miles (10–16 km), but slightly slower.

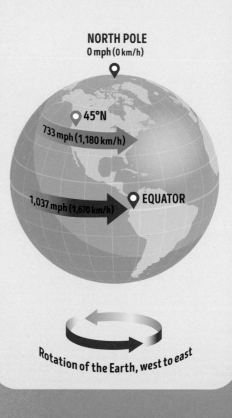

NORTH POLE
0 mph (0 km/h)

45°N
733 mph (1,180 km/h)

1,037 mph (1,670 km/h)   EQUATOR

Rotation of the Earth, west to east

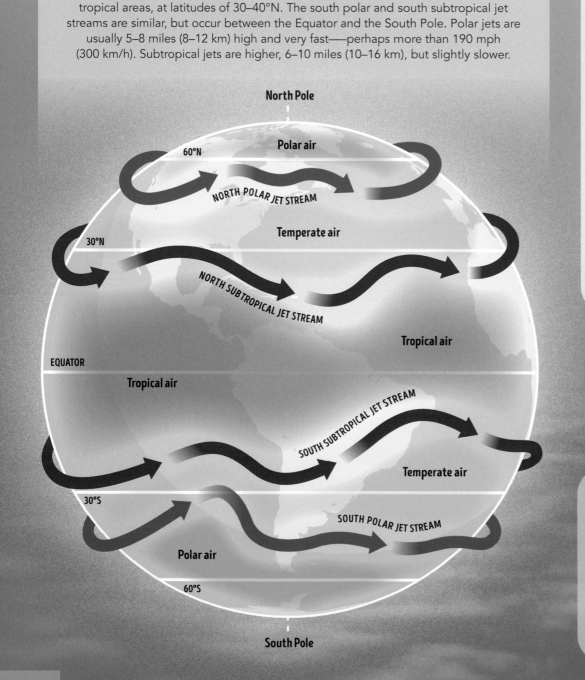

North Pole

Polar air

60°N

NORTH POLAR JET STREAM

Temperate air

30°N

NORTH SUBTROPICAL JET STREAM

Tropical air

EQUATOR

Tropical air

SOUTH SUBTROPICAL JET STREAM

Temperate air

30°S

SOUTH POLAR JET STREAM

Polar air

60°S

South Pole

JET STREAM FLIGHT TIMES

Many jet aircraft "ride the jet stream" eastward to save time and fuel. For example, if strong jet stream winds are in certain places over the North Atlantic, the journey east from North America to Europe may be two hours quicker than the same journey west, against the jet stream.

WAVY, WANDERING WINDS

Jet streams are wavy, or undulating, curving to the north and south. And these wavy curves are often on the move. Sometimes, they're curvier and at other times straighter. They may weaken for a time, then return with greater strength. Occasionally two jet streams may even merge for a while. These movements can have profound effects on weather. For example, a region to the south of the north polar jet stream generally has milder, more temperate weather. But if the jet stream curves to the south of this region, as shown here in these maps of the United Kingdom, colder polar air arrives.

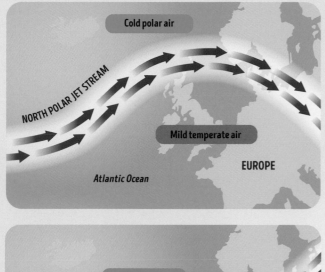

BETWEEN POLAR AND TEMPERATE

In the lands south of the Arctic Circle, but usually north of the north polar jet stream, there are vast forests, like this one in Russia, made up mainly of evergreen trees. The forests extend in a band all around the world, across North America, Northern Europe, and North Asia. These regions are known as taiga, or boreal forest. There are few roads or towns. The climate is between polar and temperate, with very cold winters, and snow and ice for perhaps half the year.

TEMPERATE CLIMATES

Temperate climates are moderate, or mild, lacking wide variations. Most are found between the jet streams, in the midlatitudes. At midday, the Sun is neither high overhead nor very low on the horizon. There are warm summers and cool winters. Because of this lack of extremes, many of the world's most extensive farmlands and big cities, like Buenos Aires in Argentina (shown here), are in lowland temperate climate zones. The main variations are in rainfall, which is affected by wind direction and distance from the coast.

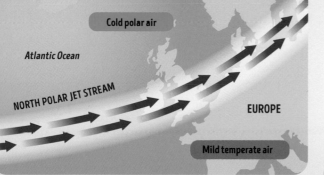

MOIST TEMPERATE CLIMATES

In many temperate midlatitude regions, the climate is warm and moist enough for trees to grow, allowing woods and forests to thrive. Some are dominated by evergreen trees, which have leaves all year. Other woodlands have mainly deciduous trees, also known as broadleaf trees. These lose their leaves for part of the year, usually in fall and winter, then grow new ones in spring. Some forests have a mixture of evergreen and deciduous trees. In a few places the climate is damp enough for temperate rainforests.

DISAPPEARING FORESTS

The climate that allows these forests to grow is also good for farming. Vast areas of temperate forests have been cut down, especially in the Northern Hemisphere. Their places have been taken by all kinds of farmland (such as these tea plantations on Angkhang mountain, Chiang Mai, Thailand), as well as by towns, cities, transportation systems, and facilities for leisure and tourism. The remaining trees often suffer from polluted air and water. Most people now understand the importance of temperate forests for wildlife and for slowing climate change. Conservation and replanting work mean that areas of temperate forests are now slowly increasing.

AMAZING FALL COLORS

In temperate areas, most deciduous, or broadleaf, trees have wide, flat leaves to take in maximum sunlight for growth. But such leaves are damaged by winter frosts and snow, so they are shed in fall. As the trees take back valuable nutrients from the leaves, and the leaves begin to break down, they change color, often into yellows, oranges, and reds. Fall in a deciduous forest is a very colorful time, as this spectacular display in the Adirondack Mountains, New York State, shows.

ANCIENT REMNANT

Białowieża Forest, on the Poland–Belarus border, is one of the largest surviving areas of Europe's original cool temperate forests. A World Heritage Site of almost 580 square miles (1,500 sq km), its continental interior climate gives it contrasting seasons. Its four months of summer have an average temperature of 68–75°F (20–24°C), but over the six months of winter this drops to 27–35°F (–3°C–1°C). An important wildlife refuge, Białowieża is the only remaining native range of the European bison (below) and is also home to deer, wild boar, bears, wolves, lynx, buzzards, and eagles.

TEMPERATE DECIDUOUS TREES

Certain kinds of trees have spread and adapted well to different temperate climates, soils, and landscapes in different regions around the world. They include oaks, elms, maples, and birches. The first oaks date back more than 50 million years. As prehistoric climates changed and shifted, oaks evolved to keep up, partly because the different species could easily breed together to form new kinds suited to new climate zones. There are now more than 500 oak species, found on all northern continents, including the Mongolian oak, shown here on the island of Shkot, Russia.

MILD AND WET

Rainforests are not only tropical. They also grow in a few temperate zones with temperatures of 41–77°F (5–25°C) and abundant rainfall of at least 80 in. (2,000 mm), both spread throughout the year. Most of these rainforests are found along western coasts in temperate latitudes. Here, between the Tropics and polar regions, prevailing winds blow from the west (see page 18). They also travel long distances across open oceans, picking up lots of moisture. In addition, the moderating effect of the ocean (see page 74) makes the winds less extreme in temperature. The result is a relatively constant, cool, wet climate all year round.

Temperate rainforest in Tasmania, Australia

CASE STUDY:

AMERICAN NORTHWEST: RAINY FOREST

One of the planet's biggest and best examples of a temperate rainforest is on North America's northwest Pacific Coast. It extends along a narrow coastal strip from Alaska, south through Canada, and down to northern California. Cool and damp for most of the year, it is world famous for containing some of the planet's oldest, tallest, and most massive trees. As well as these giant plants, the forest floor is covered with mosses, ferns, and low, lush, dripping vegetation.

LOCATION

LUSH GREEN FORESTS

Olympic National Park in Washington State receives moisture-laden winds from the Pacific Ocean. These rise over the uplands to drop plentiful rain—in some years more than 160 in. (4,000 mm). The main trees are conifers, such as firs, spruces, cedars, and hemlocks. Their year-round foliage shades the forest floor where damp-loving sedges, mosses, and ferns thrive in the cool shade. In 1981, the park was awarded World Heritage Site status.

Olympic Park's climate is made more moderate and wet by the neighboring Pacific Ocean. Average daytime highs in its rainforest are 72°F (22°C) in summer and 44°F (7°C) in winter. The rainiest month is January, with 20 in. (520 mm)—more than half the amount inland Seattle, only 72 miles (115 km) away, receives all year.

OUTSIZED TREES

The year-round cool, damp, relatively constant climate of Pacific temperate rainforests, with no heat waves or droughts, allows enormous trees to thrive. Tallest is "Hyperion," a coastal redwood in California's Redwood National and State Parks. More than 380 ft. (116 m) high and over 700 years old, it is still growing 1.6 in. (40 mm) each year.

SECRETIVE AND RARE

The mountain beaver is one of the Pacific rainforest's strangest creatures and is found nowhere else. More closely related to squirrels than its beaver cousins, it is 20 in. (50 cm) long and eats the ferns and soft plants that grow on the forest floor. Rather than build dams, it digs burrows in the loose, damp soil and can also climb trees. These creatures don't hibernate and have poor control over their body temperature, so can only thrive in the moist, temperate conditions of the Pacific rainforest, where there is rarely snow or very cold temperatures.

MEDITERRANEAN CLIMATES

The temperate climate known as Mediterranean has mild, damp winters, and dry, warm, or hot summers. It is named after the general weather conditions of the coastal areas around the Mediterranean Sea, between Europe and Africa, although similar climates in other regions of the world are also known as Mediterranean. The moderate winter temperatures, and in particular the striking contrast between rainfall in winter compared to summer, are influenced by the nearby sea and its currents.

MEDITERRANEAN SHORE

The vegetation bordering this Maltese beach is typical of Mediterranean coastal areas: sparse, tough, and drought-tolerant. The island of Malta has low annual rainfall, only 24 in. (600 mm), and no large rivers or lakes. Its water supplies come from groundwater in the rocks below the surface, and desalination —removing salt from seawater. However, these sources are limited and energy-consuming. The country is introducing many schemes for water conservation and recycling.

NOT JUST THE "MED"

Mediterranean climate regions are found in the far southwest and south of Australia, along coastal California and Baja in North America, on the northern and central coasts of Chile, and on the southwest tip of South Africa. Each has unique plants and animals specially adapted to the conditions, particularly the hot, dry summers. These landscapes and wildlife are the focus of special conservation measures in every country where they occur.

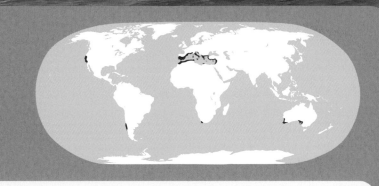

MEDITERRANEAN SCRUB AND SHRUB

Each Mediterranean climate region is characterized by a type of shrubland or scrub with grasses, bushy thickets, and occasional trees. These plants, known as sclerophylls, have tough evergreen leaves that reduce moisture loss during the long, dry summers. Although the landscapes of the various Mediterranean climate regions look similar, the actual plants differ, as do the names for this terrain:

- ☀ Matorral or matagal in Chile, Portugal, and Spain
- ☀ Maquis and garrigue in France
- ☀ Macchia Mediterranea in Italy
- ☀ Batha in the Eastern Mediterranean
- ☀ Fynbos in South Africa
- ☀ Mallee and kwongan in Australia
- ☀ Chaparral in California, USA

Mallee scrubland on the Nullarbor Plain, Western Australia

VACATIONS AND WILDLIFE

The Mediterranean coast's climate of sunny summers and usually calm seas attract millions of vacationers, especially from more northerly parts of Europe. They crowd the beaches, bars, and hotels. However, poorly controlled tourist development and increasing pollution have destroyed many natural coasts where seals, fish, shellfish, and rare plants once thrived, and where turtles came to lay eggs. In recent years, people have become more aware of these problems, with areas now set aside for wildlife.

A loggerhead sea turtle returns to the sea at sunrise, after laying eggs on Turkey's Mediterranean coast.

SEASONAL CHANGES

With their comfortable temperatures, Mediterranean climates are pleasant all through the year. Average winters are 41–68°F (5–20°C), so there are few frosts and little snow. In summer, the averages are around 68–80°F (20–27°C), with plenty of sunshine, little rain, and occasional hot spells of more than 90°F (32°C). In early summer, the countryside is alive with colorful flowers, such as these on the hills of Italy's Amalfi coast.

PRECIOUS SPECIES

Numbats are marsupials, related to kangaroos and koalas. They were once common in the Mediterranean scrublands and woods in Western and South Australia, feeding mainly on termites. But when Europeans began farming and brought foxes and cats with them from the 19th century onward, numbat numbers fell drastically. There are probably fewer than 1,000 left in just a few very small areas. They are now officially listed as an endangered species and protected.

ARID CLIMATES: DESERTS

Deserts have very dry, or arid, climates. Scientifically, they are places that receive less than 10 in. (250 mm) of rain and other precipitation in an average year. They cover about one-fifth of Earth's land area. A popular image of a desert is the Sun beating down from a cloudless sky onto sandy plains and dunes, with a few cacti and similar plants. However, only one-fifth of desert areas are sandy, and deserts can be cold as well as hot. There are many other kinds of desert landscapes, including boulder-strewn plains, rocky hills, mountains, and ice sheets.

HOT, COLD, AND COASTAL

Generally, hot deserts have an annual average temperature of over 68°F (20°C). In cold deserts, daytime maximums are rarely higher than 77°F (25°C), while minimums at night often fall well below freezing. The two main coastal deserts are the Atacama in South America and the Namib in Africa. Both are on western shores where cold currents well up from the deep ocean. Prevailing westerly winds carrying lots of water vapor are cooled as they pass over these upwellings, forming low clouds, fog, and perhaps a little drizzle. By the time the air reaches the shore it has very little water left.

Arctic Desert   Arctic Desert

Karakum
Kyzylkum   Gobi

Judaean

Great Basin

Mojave   Ordos

Sonoran   Chihuahuan   Taklamakan

Sahara   Arabian

Thar

Nubian

Denakil

TYPE OF DESERT

● Cold desert

● Hot desert   Sechura

● Coastal desert

Kalahari

Namib   Atacama

Great Sandy   Tanami

Gibson   Simpson

Pinnacles

Great Victoria

Patagonian

Antarctic Desert

CLIFFS AND CANYONS

The eastern parts of the Judaean Desert in Israel and the West Bank (left) have annual rainfall of less than 4 in. (100 mm). This is partly due to the rain shadow effect of the Samarian Hills and Judaean Mountains to the west. The nearby city of Jerusalem, only 31 miles (50 km) to the northwest, has a rainfall of about 22 in. (550 mm) yearly. The desert is noted for its numerous cliffs and canyons, some deeper than 1,150 ft. (350 m).

FOGGY DESERT

The Namib of southwest Africa is a coastal desert where there is virtually no rain at all. Parts of the shore have less than 0.4 in. (10 mm) a year. However, the area does receive some additional moisture in the form of fog drifting in from the Atlantic Ocean (left). This allows a few tough plants, such as acacia trees, to grow amid the sand dunes. Occasional animal visitors include elephants and giraffes.

DESERT CONTINENT 1: ANTARCTICA

The enormous, ice-covered continent of Antarctica is the world's largest desert. There is only 4–6 in. (100–150 mm) of precipitation each year, nearly all snow. Antarctica is also very high, windy, and cold. The conditions are so harsh that they can only support tiny creatures. The biggest creatures that are able to spend all their lives here are tiny midges and springtail bugs that are smaller than this "o."

DESERT CONTINENT 2: AUSTRALIA

The smallest continent, Australia, has the most desert area for its size. More than 10 named deserts make up one-third of the whole country. The Pinnacles, a group of stone towers in Western Australia (above), were probably shaped by loose, windblown desert sand wearing away softer limestone around the harder rock that formed from fossilized tree trunks.

◎ The Judaean Desert, Israel

DESERT SURVIVORS

Succulents, such as this spiny cactus in South America's Patagonian Desert, are plants that are specially adapted to arid climates. Cacti deal with the dryness by storing water in their fleshy, swollen stems, and their leaves are like spikes to reduce water loss and deter plant-eating animals. Other arid-adapted succulents use their leaves for water storage.

SIX LARGEST DESERTS

Name	Area million sq. mi. (million sq km)	Location
Antarctic	5.5 (14.3)	Around South Pole
Arctic	5.3 (13.7)	Around North Pole
Sahara	3.4 (9.1)	North Africa
Australian	1 (2.7)	Australia
Arabian	0.9 (2.4)	Arabian Peninsula
Gobi	0.5 (1.3)	Central Asia

CASE STUDY:

SAHARA: NOT ALL SAND

The Sahara Desert may not be quite the driest place in the world, but it is certainly the largest non-polar desert. It covers almost the same area as the entire USA or China. Less than one-quarter of the desert area is sand, so it is far from being an endless sea of wave-like dunes. The rest is varied terrain of gravelly soils, stony plains, rocky hills, salt flats, dried river and lake beds, and even mountains with occasional snow.

LOCATION

STONY AND BARREN

Stony, rocky plains and uplands strewn with gravel, scattered pebbles, and boulders cover about one-half of the Sahara. In this terrain, known as hamada, the winds have long since blown away smaller particles of dust and sand. Almost no plants grow and so hardly any animals live here.

SURVIVING ENDLESS DROUGHT

The Sahara covers one-third of Africa, stretching 3,000 miles (4,800 km) from west to east. It is dry for several reasons. The descending air of the atmospheric Hadley cell (see pages 18, 48) creates high pressure at the surface, keeping away moist air. Also, a few thousand years ago, the Sahara had grasses and bushes. Back then, people tended herds of goats, sheep, and other livestock, but these grazed away the vegetation. This meant that plants no longer released moisture into the air, which became drier, with fewer clouds. Some people today still keep animals, like this herd of cattle in Agadez, Niger, moving them regularly to allow new plants to grow.

OASIS OF LIFE

At a desert oasis, water allows plants, animals, and people to survive. The water source may be an underground river, a well dug into the ground, or a dip in the land where water oozes up from deeper rock layers that contain water. About 100 oases are dotted across the Sahara, like this one at Awbari, Libya.

RIVERS, LAKES, AND SWAMPS

In the distant past, the Sahara had a much wetter climate with rivers, swamps, and lakes filled with plant and animal life. Around 95 million years ago, it was home to the largest known land predator—the 60 ft. (18 m) long dinosaur Spinosaurus (above). It was an able swimmer and caught fish, turtles, and other water-dwellers.

GREEN TO BROWN AND BACK

- ☀ Over its long history, the Sahara's climate has regularly changed from damp to dry and back again.
- ☀ Today, yearly rainfall in about half of the Sahara is less than 0.4 in. (10 mm).
- ☀ In places, daytime temperatures average 104–113°F (40–45°C), dropping down to 59–68°F (15–20°C) in the cooler nights.
- ☀ There are so few clouds that the Sun shines for nine-tenths of daylight hours.

GRASSLANDS GALORE

Where the climate is too dry for woods and forests, but too moist for desert and dry scrub, there are great grasslands. They have various names around the world: steppe in Asia, prairie in North America, pampas in South America, savanna in East Africa, veldt in Southern Africa, and downs in Australia. However, many are no longer natural. They are grazed by farm animals or used to grow farm crops, such as domesticated cereals from the grass family, including wheat, barley, oats, rye, maize, and sorghum.

SPECTACULAR WILDLIFE

The grasses of the African savannas are so plentiful and extensive that they can support huge herds of large wild animals. Grazers include antelopes, gazelles, wildebeest, and zebras. The tallest are giraffes, and the largest are savanna elephants, with full-grown males weighing more than 8.8 tons (8 tonnes). In turn, a host of fearsome hunters prey on these herbivores (plant-eaters), including lions, leopards, cheetahs, hyenas, and jackals.

GRASSLAND CLIMATES

Natural and farmed grassland habitats cover about one-quarter of Earth's land surface. Their climates are relatively dry, with rain and other precipitation between 16 and 39 in. (400 and 1,000 mm) in a typical year. However, their temperatures vary hugely. On the American prairies, shown here in Yellowstone National Park, Wyoming, they may peak at 100°F (38°C) in summer yet plunge to –40°F (–40°C) in winter.

[⌾] Bison graze on the open plains of Yellowstone National Park, Wyoming.

NATURAL CYCLE

Long periods with little rain mean parched grasslands at the end of the dry season. This is a time for natural wildfires, usually activated by lightning, like this one in Purnululu National Park, Western Australia, in 1999. The flames consume vegetation, and animals flee for their lives. The habitat is well adapted, though. Grasses soon sprout from their deep roots, herbivores return to feast on fresh growth, and carnivores (meat-eaters) follow them. However, natural wildfires are becoming much more common, due to higher temperatures caused by climate change. People also use deliberate burning to clear unwanted vegetation and to create barriers against bigger wildfires.

TEEMING TERMITES

The most numerous creatures in most grasslands are tiny termites and ants. A big termite nest, like this one made by snouted harvest termites in South Africa, can contain perhaps half a million individuals. Their food-gathering and droppings help to recycle and spread nutrients through the soil. Ants and termites also provide food for many insectivores (insect-eaters), from beetles and spiders to lizards, birds, and mammals such as the aardvark of Africa and giant anteater of South America.

HUMAN TAKEOVER

Most natural grasslands have disappeared. Grazing for cattle, sheep, and similar livestock has taken over enormous areas, like here on the pampas in Argentina, South America. The cattle are slaughtered for steaks, burgers, and other beef products. The largest beef producers are the USA, Brazil, the European Union, China, India, and Argentina. As little as one-tenth of the world's natural grasslands are left.

CASE STUDY:

MONGOLIAN STEPPE: GRASS TO THE HORIZON

The immense and varied steppe grasslands of Eastern Europe and Asia span a vast distance from Romania, Bulgaria, and Hungary in the west to Mongolia, China, and Russia in the east—a distance of more than 4,000 miles (6,500 km). In places, there are swaying grasses and similar plants as far as you can see, yet no trees, and also no signs of human settlement or activity. In Mongolia, more than half of the land area is steppe, and around one-third of the country's population of more than 3 million lives a nomadic lifestyle on the grasslands.

ON THE MOVE

With such a dry continental climate and bitterly cold winters, even grasses grow slowly on the Mongolian steppes. People raise horses, goats, cattle, sheep, camels, and yaks, but they must follow a nomadic lifestyle, moving regularly to find fresh grazing. They use their animals to carry their dwellings —large tents called yurts or gers (shown below) —which are designed to be moved regularly, along with their belongings.

STEPPE GRASSES

Grasses, like these three examples that grow on the Mongolian steppes, have many adaptations to dry climates. Their long, thin stems and leaves are covered with a tough outer layer that slows moisture loss. In drought conditions, the leaves roll up to reduce evaporation of water even more. Their branching roots are widespread to find as much water as possible. Their flower structures—spikelets—are small, without large petals that lose water.

Feather grass

Fescue

Chinese lyme grass

HIGH AND WINDY

The steppes of Mongolia are high—more than 3,300 ft. (1,000 m) above sea level, and almost constantly windy. To the south is the even more arid expanse of the Gobi Desert. The cold, drying winds mean that any moisture from the limited rainfall on the steppe soon evaporates. To help deal with these conditions, the saiga antelope (above) has a large nasal cavity that filters dust from air breathed in during the dry summer and that warms fiercely cold air during the harsh winter.

SUITED TO THE CLIMATE

There are up to 2 million Bactrian, or two-humped, camels in Asia. These creatures are well suited to dry climates. Their humps store fat that can be broken down to provide water within the body. Their thick furry coats protect against both hot sun and freezing winds. Most Bactrians are now domestic animals, used for transportation, meat, milk, skins, and wool.

TROPICAL RAINFORESTS

Tropical rainforests receive more year-round rainfall than any other habitat on Earth. They are warm all year round, too. These climate conditions are ideal for living things that thrive in the heat and dampness. Because tropical rainforests have a greater number of plant and animal species than most other environments, they are known as biodiversity hot spots. They are the planet's natural hothouses of life.

STEAMY HEAT

Tropical rainforests, such as this one in Gunung Halimun National Park, Java, Indonesia, are warm to hot all year. In the lowlands, average daily temperatures range from 68–86°F (20–30°C) and rarely fall below 60°F (15°C) even on the coldest nights. Rainfall generally exceeds 80 in. (2,000 mm) each year and can sometimes be more than 200 in. (5,000 mm).

MAJOR TROPICAL RAINFORESTS

Tropical rainforests are found between the Tropics of Cancer and Capricorn. These regions are warm because the midday Sun shines almost directly above, rather than at a low angle. They are wet because of atmospheric circulations called Hadley cells on either side of the Equator (see page 18). These create an area of low pressure where air freely rises, drawing in surrounding moist air that condenses into clouds. The forest trees create their own regional climate, too. They release enormous quantities of water vapor by transpiration (see page 27), which rises, condenses into clouds, and returns to Earth as rain.

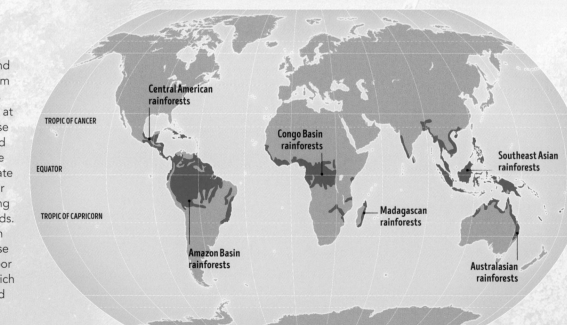

Central American rainforests

TROPIC OF CANCER

Congo Basin rainforests

Southeast Asian rainforests

EQUATOR

Madagascan rainforests

TROPIC OF CAPRICORN

Amazon Basin rainforests

Australasian rainforests

LAYERS OF THE RAINFOREST

In most tropical rainforests, different kinds of animals and plants are found living at different levels. The forest floor is usually shady and very damp, with mosses, ferns, and low undergrowth. The rising tree trunks are draped with creepers and vines, frequented by lizards and monkeys. The sunlit treetops form a thick green layer called the canopy, packed with leaves, blossoms, and fruits. Most animal life is here, from butterflies to brightly colored birds, such as the *Eclectus roratus* parrot of northeastern Australia (left).

An aerial view of the Amazon Rainforest, Brazil

RAIN INTO RIVERS

The Amazon Rainforest climate feeds water into the world's largest river, the Amazon. At more than 4,000 miles (6,400 km) in length, the Amazon carries more water than the next six biggest rivers added together.

COOL AND CLOUDY

In hilly and mountainous areas, tropical cloud (or montane) forests grow. These are often even wetter than lowland rainforests, but chillier, too. As warm, damp air flows up the slopes, it cools and its water vapor condenses into thick clouds and rain. Temperatures are lower, not only because of the greater height, but also because the clouds and mists often obscure the Sun. Most tropical cloud forests, like the one shown here in Monteverde, Costa Rica, are at altitudes of 2,600–8,200 ft. (800–2,500 m), with average temperatures around 50–70°F (10–20°C).

MOST ANCIENT FOREST

The rainforests of northeast Queensland, Australia, are some of the oldest on the planet. The climate and vegetation here date back almost 200 million years. Lots of rain means that large rivers flow for most of the year, supporting a huge variety of life. This plentiful food supply attracts one of the biggest and oldest kinds of reptiles on the planet—the saltwater crocodile (above), which can reach lengths of 20 ft. (6 m).

CASE STUDY:

CONGO BASIN: SOAKED AND STEAMY

Second only in area to South America's Amazon, the tropical rainforests of the Congo in West Africa cover almost 1.5 million square miles (4 million sq km). The immense Congo River receives water from smaller rivers and streams spread over an area—known as its drainage basin—that covers more than 10 countries. The whole Congo Basin covers wide areas north and south of the Equator, so there is always rain somewhere to feed the river's flow.

GREATEST FALLS

The Congo River's Livingstone Falls is a series of rapids around 220 miles (350 km) in length, of which Inga Falls (below) is the largest. Due to the region's high rainfall and large drainage basin, the foaming, roaring currents carry more water than the world's next 10 largest waterfalls combined. Rather than one tall waterfall, Inga is more a series of smaller cascades and rapids with a total descent of almost 330 ft. (100 m).

GAPS IN THE RAINFOREST

Scattered here and there in the Congo's rainforests are natural wetland clearings, known as bais. Here, natural pools form out in the open. These places teem with wildlife, from fish, frogs, and wading birds to crocodiles, pythons, lowland gorillas, buffalo, and even African forest elephants. As the larger herbivores wallow, graze, socialize, and dig for nutrients, they help to keep bais clear of trees.

African forest elephants graze at a bai in Dzanga National Park, Central African Republic.

CONGO CLIMATE

- Across much of the Congo Basin there are two rainy seasons, generally March to May and September to November. Between these are "drier" seasons when the weather is not quite so wet.
- Average yearly rainfall varies from 50–80 in. (1,300–2,000 mm).
- Almost one-third of water vapor given off by the Congo's rainforest trees is recycled, forming clouds in the region and returning back to the forest as rain.
- Daily temperatures are usually in the range of 68–81°F (20–27°C).

VITAL WATERWAYS

Travel and transportation are difficult through the thick Congo rainforests. So, the rivers and lakes provide essential highways for people and cargoes. But travelers must take care. The Congo is the world's deepest river, with a depth of more than 650 ft. (200 m) in places. And there are also hidden mud banks, floating logs, and other dangers to watch out for.

MONSOON CLIMATES

The word "monsoon" is said to come from an Arabic word *mausim*, meaning "season," "windy season," or "shifting winds." It describes climates where winds, which blow in one direction for much of the year, reverse (or shift) direction, accompanied by great changes in rainfall. The largest and most powerful monsoon climates occur in South Asia—in Sri Lanka, India, Pakistan, and Bangladesh—and parts of Southeast Asia. Smaller, less marked monsoon-type climates occur in West Africa, Australia, and southwest North America, Central America, and northern South America.

WAITING FOR THE RAIN

In South Asia, the main monsoon (wet and rainy) season is usually in summer from late May to September. The rainfall patterns spread from the southwest, arriving first in Sri Lanka, then spreading across the Bay of Bengal into Myanmar (Burma) and Thailand. In early June, they extend to the states in southern India, then east to south China. By July, the rains have usually reached northwest India and Pakistan. However, dates vary from year to year. Weather experts track the rain's path in great detail, since people need time to prepare for the deluge to come.

WIND REVERSAL

As well as the main wet monsoon season from June to September (see above), South Asia also has a second "dry" monsoon season from October to April or May. The wet monsoon is caused by the combined effects of the jet streams, atmospheric air cells, and the Inter-Tropical Convergence Zone (see page 19), which bring southwest winds laden with moisture from the Indian Ocean. These drop huge amounts of rain over the land. During the winter, or dry, monsoon season, winds come overland from the opposite direction (northeast), and so carry very little rain. In southeast India and Sri Lanka, there is also some "inter-monsoon" rainfall during late October and November.

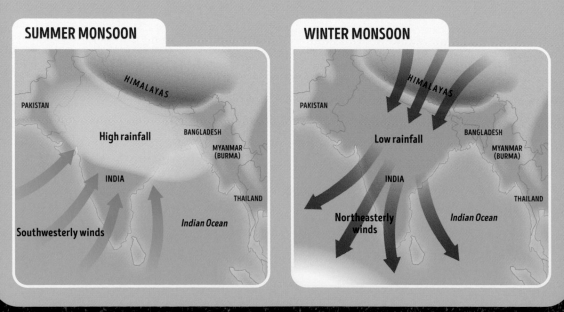

SUMMER MONSOON

HIMALAYAS

PAKISTAN

High rainfall

BANGLADESH

MYANMAR (BURMA)

INDIA

THAILAND

Indian Ocean

Southwesterly winds

WINTER MONSOON

HIMALAYAS

PAKISTAN

Low rainfall

BANGLADESH

MYANMAR (BURMA)

INDIA

THAILAND

Indian Ocean

Northeasterly winds

FOOD FOR BILLIONS

Monsoon rains may cause problems, but they are vital to irrigate, or water, the land for growing crops such as rice, cotton, sugarcane, and soybean. Rice is a cereal —a member of the grass family. Grown in warmer climates across much of Asia, it is the main bulk food for about half the people in the entire world.

Harvesting rice in Vietnam in September, just after the monsoon season

FLOOD DISASTERS

Most monsoon climate regions cope with the usual heavy rainfall, which can exceed 40 in. (1,000 mm) in a few months. The water is channeled into crop fields, drains, ditches, rivers, and reservoirs. However, an especially heavy monsoon may bring large-scale destruction. In Thailand, the 2011 monsoon floods devastated farmland, industries, transportation systems, towns, and cities, as shown above in the city of Nonthaburi. More than 800 people died. It was among the top 10 most costly natural disasters of all time.

◎ Monsoon rain hammers down in Kerala, southern India.

CASE STUDY:

FARMING IN BANGLADESH

The monsoon climate of South Asia is ideal for growing one of the world's staple foods—rice. More than two-thirds of land in Bangladesh is used for farming and over two-thirds of this farmland is planted with rice. The country's monsoon rains are crucial to the crop because rice is very "thirsty," requiring lots of water to thrive. Different kinds of rice have been selected and bred over thousands of years to grow at different times throughout the year, mainly in semi-flooded fields called paddies. Bangladesh also has a new, fast-growing trend—aquaculture, or fish-farming— which is also made possible by its monsoon climate.

LOCATION

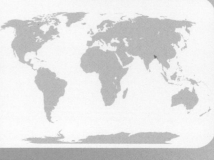

SEASONS OF RICE

Bangladesh's climate allows three main rice growing seasons, keeping the land and its farmers busy throughout most of the year. These seasons are called *aus*, *aman*, and *boro*. The aus season rice crop is planted during March and April, toward the end of the dry monsoon period, and harvested from June to August. The aman season planting begins just before, or at the start of, the wet monsoon period, from June, and is harvested in November. The boro season rice is planted in November and December, soon after the dry monsoon period begins, and is harvested toward the end of this season, during April and May. At all times, the growing rice needs plenty of water, either from fresh rainfall or stored supplies.

 Rural farmers in Bangladesh plant out rice seedlings in the flooded paddy fields during the aman season.

SHIFTING CROP AREAS

On World Food Day, October 16, 2021, the Bangladesh government announced the country had achieved its long-term aim of having become largely self-sufficient in food. However, it also stated that further improvements were needed to match crops more precisely to the climates of different regions. For example, rice is best suited to the high, relatively reliable rainfall in eastern regions. Vegetables and fruits are more suited to central and west where the rainfall is more variable from year to year. This variety means a wider range of home-grown produce in local markets, as shown here in Dhaka, Bangladesh.

FISH FARMING

Bangladesh's plentiful supply of water from its monsoon climate is also being exploited to raise freshwater fish, shellfish, and even water snails, especially in the north. The number of freshwater fish ponds, lakes, and similar enclosures in the country exceeds 2 million and is growing fast. Saltwater fish farms in coastal areas are also on the increase. The nation is now one of the world's top five aquaculture producers.

Freshwater fish farming in Bangladesh

WHEN MONSOONS FAIL

If monsoon rains do not arrive, the rice crop is much reduced and there are widespread food shortages, not just in Bangladesh but also in nearby countries. This happened in 1997 and again in 2009. Scientists linked these episodes to an increase in extreme weather events due to climate change. In 1998 and 2019, extra-heavy monsoons caused widespread floods that washed away rice and other crops.

CONTRASTING SEASONS

The port city of Barisal on the Kirtankhola River lies in one of the country's biggest rice-growing regions. In a typical year, the area receives more than 80 in. (2,000 mm) of rain. More than four-fifths falls in the wet monsoon between May and September. In contrast, the daytime temperatures vary by only a few degrees between the wet and dry seasons.

MONTH	RAINFALL		TEMPERATURE (high)	
	in.	mm	°F	°C
January	0.4	10	77	25
February	1	26	82	28
March	2	52	90	32
April	4	104	93	34
May	7.8	200	91	33
June	15.8	402	90	32
July	16	410	88	31
August	13.5	13.5	88	31
September	11.2	285	90	32
October	7.3	185	90	32
November	2	50	86	30
December	0.2	6	80	27

MOUNTAIN AND UPLAND CLIMATES

When it comes to weather and climate, higher altitude means colder temperatures. Average air temperature drops by about 3.5°F per 1,000 ft. (6.5°C per 1,000 m) of altitude. So the top of a mountain 16,500 ft. above sea level is about 59°F (33°C) colder than nearby land at sea level. Higher altitude also means more wind, since moving air is not slowed by friction with the ground (see pages 14–15). These conditions give hills, mountains, and other highlands a very different climate to lowland areas. Different plants and animals live here, too.

CHANGING SCENERY

Mountain ranges in temperate zones, such as the Alps in Europe (below), have spectacular landscapes—and also show the effects of altitude. In lower, milder areas, there are grassy fields and flowers. Farther up, in cooler, damper areas, deciduous, or broadleaf, trees grow. With increasing altitude, these gradually give way to conifers and other evergreens. Even higher up, it becomes too cold and windy for trees. There may be small, stunted bushes and other tough, low-growing plants, but ice and snow lie on the ground for much of the year.

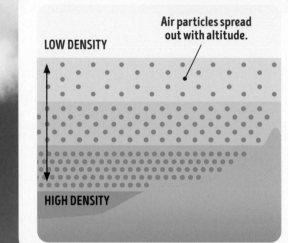

Air particles spread out with altitude.

LOW DENSITY

HIGH DENSITY

LESS AIR, LESS WARMTH

Temperatures fall with altitude partly because the atmosphere becomes thinner. This means there are fewer molecules of air to hold on to any heat. And because they are fewer and farther apart, these molecules bump into each other less as they move about, which also generates less warmth. In addition, heat from the Sun that is reflected by the sea and land only reaches (and warms) the lower levels of the atmosphere.

SKI'N'SURF

In a few places around the world, people can ski and surf on the same day, due to a happy combination of altitude and climate. High on mountains near the coast, it is cold enough to ski or snowboard. Then, at sea level, just a couple of hours' travel away, there are warm beaches and balmy seas. Such locations include California's Big Bear ski resorts and the beaches of Los Angeles (right), which are just 100 miles (160 km) apart but have an altitude difference of around 8,860 ft. (2,700 m). Other ski'n'surf locations include the Pyrenees Mountains and Biarritz in France, the west coast of New Zealand's North Island, and the Atlas Mountains and Atlantic Coast of Morocco.

Big Bear Mountain, California

Malibu Beach, Los Angeles, California

AT THE TOP OF THE WORLD

Places with a higher altitude have generally colder climates, but their location on Earth—whether near the icy poles or sweltering Equator—also has a big influence. At the world's highest point—the summit of Mount Everest on the China–Nepal border, 29,032 ft. (8,848 m) high—temperatures plummet to -76°F (-60°C). However, Everest is only 320 miles (520 km) north of the Tropics. If it was at the South Pole, its temperature would be an estimated 45–55°F (25–30°C) colder. In 2019, a weather recording station (left) was set up just 1,400 ft. (420 m) below Everest's summit to monitor conditions.

HIGH-ALTITUDE WILDLIFE

Cold, windy, mountain climates greatly limit wildlife. Mammals need thick fur and birds grow dense plumage to keep out the wind and cold, and to keep in their body warmth. Snow leopards (right) roam the Himalayas and other ranges in Central Asia. They have thick, dense fur, small ears to reduce heat loss, and wide paws that work like snowshoes. To follow their prey, such as sheep, deer, and goats, snow leopards move up and down the slopes with the seasons. They descend to almost 10,000 ft. (3,000 m) or below in winter, and move to 16,400 ft. (5,000 m,) or even higher, in summer.

COASTAL CLIMATES

The Sun's heat warms land faster than water (see page 42). However, seas and oceans store their heat better than land, partly because their currents carry the warmth around and spread it out. They act as heat reservoirs, both warming up and cooling down more slowly than land. This makes climates along coasts more moderate, or less extreme, than inland regions. Also known as oceanic, or maritime, climates, coastal climates typically have rainfall spread evenly throughout the year. Their effects can be felt hundreds of miles inland.

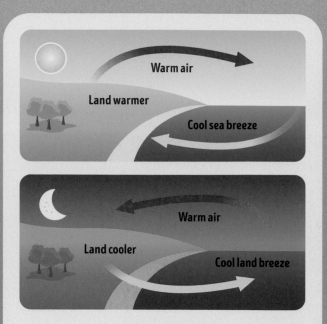

HEAT TRANSPORTER

In the North Atlantic, several huge ocean currents carry warmer surface waters from the Tropics northward. One is the Gulf Stream (see page 21). Its flow brings heat from the Gulf of Mexico, Florida, and the Caribbean northeastward to the west coasts of Britain and northwest Europe. This has a moderating effect on temperatures, meaning various kinds of warmth-loving plants can thrive in more northerly climes, as seen here in Torquay, southwest England. Without the Gulf Stream, these coasts would be perhaps 14°F (8°C) colder, while inland would be 9°F (5°C) cooler.

COASTAL BREEZES

During the daytime, the Sun can warm land by 18°F (10°C) or more, while the nearby sea warms by just 1.8°F (1°C) or less. Air heated by the land rises, and cooler air from the sea flows in. This wind, from the sea onto the shore, is a sea, or onshore, breeze. At night, the pattern reverses as the land cools more quickly. The air over the sea is now warmer and rises, drawing air from the land out to sea as a land, or offshore, breeze. People who do watersports, such as kitesurfers, sailboarders, and surfers, use these winds to maximum effect.

SEA FOGS

Sea fogs drifting in from the ocean are common in some coastal areas. Japan's Pacific shores—especially southeast Hokkaido (below) and northeast Honshu islands—experience numerous sea fogs, particularly in spring and summer. These are due to warm, moist air, cooled by cold ocean currents, being blown onto the land by easterly winds. Sometimes the fog extends across the Pacific for hundreds of miles.

THE SAME BUT DIFFERENT

The city Punta Arenas in southern Chile, near the tip of South America, is one of the largest cities on Earth to be located so far south—only 2,500 miles (4,000 km) from the South Pole. However, its close proximity to the Pacific Ocean makes its coastal climate relatively even throughout the year, both in temperature and rainfall. In comparison, Winnipeg in Canada is about the same distance north of the Equator as Punta Arenas is south. But Winnipeg is far from both the Pacific and Atlantic Oceans. As a result, its continental climate is much more extreme, with warm, damp summers and long, dry, freezing winters.

	PUNTA ARENAS	WINNIPEG
MIDSUMMER – daily average temperature	50°F (10°C)	68°F (20°C)
MIDWINTER – daily average temperature	35°F (2°C)	3°F (-16°C)
MIDSUMMER – monthly average precipitation	1.2 in. (30 mm)	3.5 in. (90 mm)
MIDWINTER – monthly average precipitation	1.7 in. (43 mm)	0.8 in. (21 mm)

Punta Arenas, Chile, enjoys a surprisingly mild climate for its southerly location.

ISLAND CLIMATES

Small mid-oceanic islands have climates dominated by the wind, waves, and water currents around them. The smaller and flatter the island, the more the weather is like the surrounding ocean. Larger islands with hills and mountains, however, create their own microclimate zones. For example, on the main wind-facing, or windward, side of an island, when moisture-laden oceanic winds blow up slopes, they cool to form clouds, then release rain. This means that the upland areas are much wetter than the nearby lowlands.

WEST PACIFIC EQUATOR: WARM AND WET

Like the Galápagos (lower right), the tiny island nation of Nauru is almost directly on the Equator. However, a distance of 7,150 miles (11,500 km) separates these two locations—more than one-quarter of the distance around the Earth. Nauru's local ocean currents and winds are warm and wet, and the island has a partly monsoon climate. Its coral reefs (below) throng with colorful life.

EAST PACIFIC EQUATOR: COOL AND DAMP

Like Nauru (below left), the Galápagos Islands are on the Equator and should be very warm. But especially from June to December, a powerful current brings cold water from the south, and along with cool southern trade winds (see page 18), this makes the sea and air cooler. Moist ocean air rises up the hilly slopes and its vapor condenses, shrouding the lands in mist and causing a lack of sunshine that adds to lower daytime temperatures.

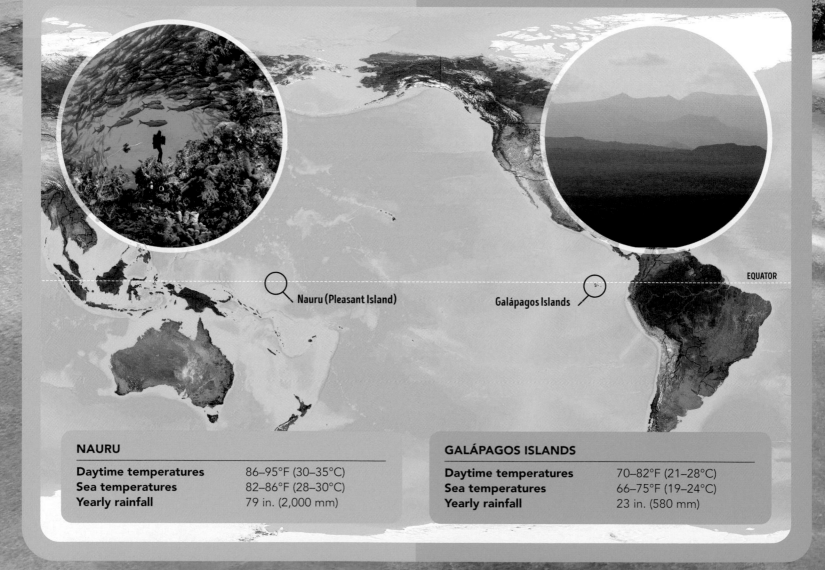

Nauru (Pleasant Island)

Galápagos Islands

EQUATOR

NAURU	
Daytime temperatures	86–95°F (30–35°C)
Sea temperatures	82–86°F (28–30°C)
Yearly rainfall	79 in. (2,000 mm)

GALÁPAGOS ISLANDS	
Daytime temperatures	70–82°F (21–28°C)
Sea temperatures	66–75°F (19–24°C)
Yearly rainfall	23 in. (580 mm)

📷 The low-lying Marshall Islands, surrounded by the western Pacific Ocean

SUNNY VACATIONS

The Spanish Canary Islands in the eastern Atlantic, about 100 miles (160 km) from northwest Africa, have a famously warm, dry, sunny climate that attracts more than 10 million visitors each year. Annual rainfall in the driest eastern areas is only 6 in. (150 mm), which qualifies as desert, while the damper parts to the west receive 20 in. (500 mm). Daytime temperatures range from about 68°F (20°C) in winter to almost 86°F (30°C) in summer, and average sunshine approaches eight hours per day. Northeast trade winds (see page 18) can be very strong, making the Canaries a global center for kitesurfing, windsurfing, and sailing.

NOT QUITE SO ICY LAND

In the North Atlantic, Iceland is only slightly south of the Arctic Circle, and a popular location to see the Northern Lights, or Aurora Borealis (above, and see page 45). The warm waters of the Gulf Stream and other currents make its climate moderately mild, especially along the south and west coasts. Being relatively large, 305 miles (490 km) west to east and 192 miles (310 km) north to south, Iceland has several climate zones. The climate of the capital Reykjavík, in the southwest, is known as subpolar oceanic (see below). Daytime temperatures range from 37°F (3°C) in winter to 59°F (15°C) in summer. The central uplands are usually about 18°F (10°C) colder and also drier, with a tundra-type climate.

COMPLICATED CLIMATES

The main system for describing climates is the Köppen Climate Classification (see page 42). It uses names and code letters to describe various types of climates. Climatologists and meteorologists use this system as a quick, concise way of describing a climate that would take much longer to write or describe in words. For example, a temperate oceanic or marine climate is classed as Cfb.

☀ C is the temperate group of climates (A being tropical, B arid, D continental, and E polar).
☀ f indicates no particularly dry season (w is dry winter, s is dry summer).
☀ b means warm summers (a is for hot summers and c for cool summers).

Cfb climates apply mainly to islands and coasts, but they can also extend hundreds of miles inland, even to the Himalayas in Asia and parts of Central Africa.

CASE STUDY:

REMOTE ISLANDS: BOUVET AND RAPA NUI

The island that is farthest from any other land on our planet is Bouvet Island in the South Atlantic Ocean. Rapa Nui (Easter Island) in the South Pacific Ocean is almost equally remote. For both, the closest mainlands are more than 1,000 miles (1,600 km) away. This means these islands have no nearby large landmasses to influence their climate. Both islands are also very small with little surface area to create their own microclimates. The result is that each island mostly experiences the climate of its encircling ocean. However, the distance from the Equator makes their climates very different.

LOCATION

Rapa Nui

Bouvet Island

STATUES REPLACE TREES

People probably first set up home on Rapa Nui in the 12th century. At that time, there were palms and other trees, wildflowers, and various animals including several kinds of flightless birds. The settlers gradually cut down the trees, and without their shelter and nourishment, much of the wildlife soon disappeared. While the inhabitants were destroying their environment, they fashioned huge head-and-shoulder statues, known as moais (shown here).

LONG WAY TO ANYWHERE

Bouvet Island has no close neighbors. The nearest land of any kind is Antarctica, 1,050 miles (1,700 km) away, although this is uninhabited. The nearest inhabited mainland is much farther away—1,620 miles (2,600 km)—in South Africa. Rapa Nui's nearest neighbors are small Pacific islands, 1,620 miles (2,600 km) away. The closest inhabited mainland is Chile in South America, 2,240 miles (3,600 km) away.

COPING WITH THE OCEAN

Measuring just 16 miles by 7.5 miles (25 km by 12 km), Rapa Nui is 1,800 miles (3,000 km) south of the Equator. Bouvet Island (below and right) is about half this size and twice as far from the Equator, at around 3,725 miles (6,000 km) south. Such specks of land in such vast oceans are very exposed, making their island climates difficult for plants, animals, and people to thrive. Bouvet Island has no permanent human settlement, but it does have a research center and a weather station that helps to forecast storms in the South Atlantic.

A NEW RAINFOREST?

Rapa Nui is closer to the Equator than Bouvet Island, so its climate is milder. Summer daytime averages are 73°F (23°C), and in winter 64°F (18°C). Its plentiful yearly rainfall of 49 in. (1,250 mm) is similar to that of a subtropical rainforest or moist forest climate, so it is wet enough for trees to grow. However, Rapa Nui had no trees when European explorers first arrived in 1772—the original islanders had cut them down over the preceding centuries. Replanting with thousands of seedlings and saplings is now underway (below).

MARITIME ANTARCTIC CLIMATE

Bouvet Island is often shrouded in fog and mist, with low clouds that block the Sun's warmth. Its temperatures are remarkably constant through the year, with daily averages of 36°F (2°C) in midsummer and 27°F (-3°C) in midwinter. Annual rainfall is 27.5–29.5 in. (700–750 mm). Nine-tenths of the land is glacier-covered. Only a few simple plants and tiny creatures survive here. The main animals are visiting seabirds and seals.

EXTREME WEATHER

Extreme weather can cause huge amounts of damage. Major weather events, such as floods, droughts, gales, blizzards, and heat waves, can knock down buildings, ruin farmland, and wreck lives.

Dangerous weather comes in many forms. On a small scale, a sudden local downpour can wash away a farm's crops and flood roads in nearby towns. Bigger incidents disrupt whole cities for many days, or even weeks. Over months and years, droughts gradually damage plant life, so that people and animals go hungry. Some of the most awe-inspiring weather events are hurricanes and typhoons. In just a few hours they can wipe out vast areas, destroy nature and agriculture, and seriously damage urban areas. Repairs can cost billions. The problem we face is that, due to global warming and climate change, these events are becoming more common — and even more extreme.

HOW MUCH EXTREME WEATHER?

Experts say that in the past 50 years, extreme weather events around the world have become four or five times more common. That means there is one weather disaster, somewhere, every two days. In the next 50 years, if the climate continues to warm, this increase will speed up. There could be several big weather catastrophes every day.

A BLAST IS COMING

In the Southwest of the USA, a fierce dust storm, or haboob, races across the Arizona desert (main picture). It sandblasts everything with stinging windblown grains, and leaves behind a suffocating blanket that smothers plants and chokes animals. The next rain to wash away the sand might not be for months.

HOT AND COLD

Apart from rainfall, temperature is probably the most discussed aspect of weather and climate. Where are the hottest and coldest places and when did they break the records? Are there likely to be even more extreme temperatures soon? Heat and cold not only determine landscape and wildlife, they also affect where we live, our daily lives, and even the energy we use—for heating when it's chilly to air-conditioning in a hot spell. Prolonged icy periods or heat waves can even kill.

IS IT OFFICIAL?

To be official and certified, record-breaking weather events need approval from the World Meteorological Organization, part of the United Nations (see page 135). The WMO sets the standards and instructions for how accurate, reliable measurements are made, the equipment used, where weather stations are located, and how the readings are checked and verified.

10 HOT COUNTRIES

Identifying the world's warmest countries involves taking high, low, and average temperatures during the day and night, month by month, at various sites across an entire nation for many years. Measuring and calculating this mass of data in different ways gives different results. (Also, the results may well be estimated in a selective way— for example by the country's tourist industry.) Countries with yearly averages above 82°F (28°C) include, in alphabetical order:

- ☀ Bahrain, Middle East
- ☀ Burkina Faso, West Africa
- ☀ Djibouti, East Africa
- ☀ Kiribati, Central Pacific Ocean
- ☀ Mali, West Africa
- ☀ Mauritania, West Africa
- ☀ Qatar, Middle East
- ☀ Palau, West Pacific Ocean
- ☀ Senegal, West Africa
- ☀ Tuvalu, West Pacific Ocean

● COLD COUNTRIES   ● WARM COUNTRIES

Canada • Norway • Sweden • Iceland • Finland • Russia • Mongolia • Kazakhstan • Kyrgyzstan • Bahrain • Qatar • Mauritania • Mali • Senegal • Burkina Faso • Djibouti • Palau • Kiribati • Tuvalu

Antarctica

10 COLD COUNTRIES

Like the hottest nations (see left), there are several ways of calculating the world's coldest counties. Some nations take average temperatures across their whole area for 10 or 20 years or more. Some select a certain series of extra-cold years and use those, leaving out warmer ones. For other claims, it is only for a certain region within the country, like Siberia in Russia. Using combinations of these methods, the nations with most days in the year below freezing, 32°F (0°C), include, in alphabetic order:

- ✳ Antarctica
  (not a nation, but governed by international treaty)
- ✳ Canada, North America
- ✳ Finland, Northern Europe
- ✳ Iceland, North Atlantic
- ✳ Kazakhstan, Central Asia
- ✳ Kyrgyzstan, Central Asia
- ✳ Mongolia, Central Asia
- ✳ Norway, Northern Europe
- ✳ Russia, Northern Europe/Asia
- ✳ Sweden, Northern Europe

HOTTEST-EVER TEMPERATURE

Death Valley in southeastern California (below) is notorious for very high, sometimes deadly temperatures. The world's highest-ever air temperature was recorded here in 1913. The hottest ground surface temperature so far recorded was also taken here: 201°F (93.9°C) in 1972. The valley's name is not an exaggeration. Dozens of people have died in the area over the years, usually from heatstroke and dehydration.

FORMER RECORD-HOLDER

In 1922, a record temperature of 136.0°F (57.8°C) was detected in 'Aziziya, northwest Libya, North Africa. For many years this held the title as highest air surface temperature on Earth. However, experts began to suspect that the measuring equipment, or the way it was read, was faulty. There were no other temperatures so high in the area at the time. In 2012, they decided the reading was an error and the record for the hottest place reverted to the former title-holder, Death Valley, California.

COLDEST PLACES

Antarctica holds many weather records. It's not just the coldest continent, but also—on average—the highest, driest, and windiest one, too. The lowest-ever official air temperature on Earth, -128.6°F (-89.2°C), was detected here in 1983 at Vostok Research Station (below), 800 miles (1,300 km) from the South Pole. In 2010, weather satellites recorded an even lower -135.8°F (-93.2°C) near Dome Fuji Research Station, nearly 860 miles (1,400 km) from the South Pole. However, this result relied on remote sensing from space and was not a directly recorded air temperature, and so it's not considered an official record.

RECORD HIGHS AND LOWS AROUND THE WORLD

Location	Temperature	
Death Valley, USA	134.1°F (56.7°C)	122°F (50°C)
Riyadh, Saudi Arabia	117°F (47.2°C)	
Sydney, Australia	114.5°F (45.8°C)	
Rio de Janeiro, Brazil	109°F (43°C)	
Paris, France	108.6°F (42.6°C)	
New York City, USA	106°F (41.1°C)	32°F (0°C)
Christchurch, New Zealand	19°F (-7.2°C)	
Edinburgh, UK	6.8°F (-14°C)	
Helsinki, Finland	-29.7°F (-34.3°C)	
Moscow, Russia	-44°F (-42.2°C)	
Yellowknife, Canada	-51.2°C (-60°F)	-58°F (-50°C)
Vostok Base, Antarctica	-128.6°F (-89.2°C)	-130°F (-90°C)

SUNSHINE AND SUNLESS

Unsurprisingly, most of the world's sunniest places are in hot desert areas, especially in North America, Africa, and Australia. They receive more hours of direct, unbroken sunshine than anywhere else on Earth. Most of the least sunny places (usually because of cloud cover) are in the cold lands of the far north and south. However, there are only a limited number of sites with officially approved weather stations—so there could be several sunnier or cloudier places that are as yet unrecognized.

EVERYWHERE EQUAL?

Daylight is not the same as sunshine. Over one year, all locations on Earth receive much the same amount of daylight—that is, hours when the Sun is at, or above, the horizon. The poles have six months' continual daylight followed by six months' unbroken darkness. The Equator has 12 daylight hours every day. Places in between gain hours in summer but lose them in winter. There are some variations, however. For example, there is more daytime on mountain tops and less in deep valleys. There are also slight differences due to the Earth's orbit, shape, and atmosphere. For example, the atmosphere bends, or refracts, light so that in some conditions the Sun is still partly visible briefly above the horizon, like a mirage (see page 104), even when it has actually just dipped below the horizon.

AMOUNT OF DAYLIGHT
FROM NORTH TO SOUTH

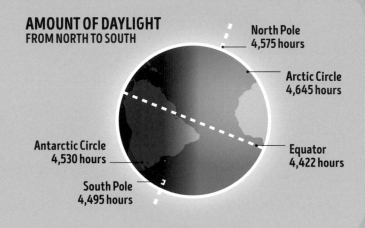

North Pole
4,575 hours

Arctic Circle
4,645 hours

Antarctic Circle
4,530 hours

South Pole
4,495 hours

Equator
4,422 hours

SOMETIMES TOO SUNNY

Official records show that the world's sunniest place is Yuma, Arizona (shown here). It has more than 4,000 hours of direct sunshine per year, averaging 11 hours daily. However there are drawbacks. Yuma is in the Sonoran Desert. Temperatures exceed 104°F (40°C) on about 100 days per year. Rainfall is less than 8 in. (200 mm). Sometimes a cool, overcast rainy day in Yuma is very welcome!

OTHER EXTREMELY SUNNY LOCATIONS

LOCATION	YEARLY AVERAGE
Marsa Alam, Egypt	3,960 hrs.
Phoenix, Arizona, USA	3,870 hrs.
Keetmanshoop, Namibia	3,870 hrs.
Aswan, Egypt	3,860 hrs.
Las Vegas, Nevada, USA	3,825 hrs.
Tucson, Arizona, USA	3,805 hrs.

ANOTHER DULL DAY

At Bear Island, part of the Svalbard group in the Arctic Ocean, the sun shines for fewer than 600 hours each year. It is cold, too, with an average daytime high of 45°F (7°C) in midsummer and 25°F (-4°C) in winter. Near the other end of the world, Motu Ihupuku (Campbell Island), far to the south of New Zealand, registers about 650 sunshine hours. Among towns and cities, Tórshavn in the Faroe Islands in the North Atlantic Ocean (below) enjoys only 840 hours of sunshine yearly. Only seven of those occur during December.

ALL DONE BY MIRRORS

One of the world's darkest towns is Rjukan in southern Norway (above). Its location in a deep valley surrounded by mountains means that its sunrise is late and its sunset early compared to flatter areas. This shortens the town's daylight hours, which are already brief in winter. In 2013, three large sun-tracking mirrors were installed on a mountainside, 1,475 ft. (450 m) above the town. Computer-controlled motors move the mirrors so they reflect the Sun's available rays down onto the town square. Otherwise it would be sunless here from the end of September until mid-March.

MEASURING SUNSHINE

What exactly is direct, unbroken sunshine? It's exactly that—a period when the Sun's rays are hitting the Earth without any interference, so measurements don't count any time when there is a faint haze in the air, or the Sun is emerging from behind clouds. Invented in 1853 by the Scottish writer John Francis Campbell, the device on the right is known as a sunshine recorder and was an early attempt to measure this. It focuses the Sun's rays through a glass ball to create a small, hot, bright point on a curved piece of cardboard. Only clear, unbroken sunshine is powerful enough to make a scorch mark on the cardboard, which gets longer as the Sun moves across the sky. Modern weather recordings use electronic sensors for this task (see page 134).

WETTEST AND DRIEST

There are some extremely wet, rainy places around the world, and also some extremely dry, arid areas. But when measuring them, it's important to take note of the timescale. For example, selecting certain years to calculate averages can produce differing results. Also, the title for record rainfall over one year might not go to the same place as the record for rainfall in one month, a week, a day, or a minute. Aridity is slightly less complex, usually because the precipitation in very dry places tends to vary less from year to year. These two extreme weather features are closely linked to record-breaking floods and droughts (see pages 92, 98).

MOST RAIN OVER THE YEAR

There are several claimants for the title of "world's wettest place" (as measured by average rainfall over a year), including the towns of Mawsynram in the hills of northeast India (shown here), and López de Micay and Lloró, both in the tropical rainforest zone of northwest Colombia. Direct comparisons are tricky because their averages are taken over different sets of years.

WHO HOLDS THE TITLE?

○ Mawsynram is often given the title of world's wettest place. Its yearly average rainfall from 1960 to 2010 is about 508 in. (12,900 mm). However, including certain decades before and after reduce the average annual amount to 467.4 in. (11,873 mm).

○ Lloró has claimed rainfall of about 500 in. (12,700 mm) between 1952 and 1990, but other averages for different years are higher, or as low as 315 in. (8,000 mm).

○ López de Micay rivals Mawsynram, although depending on the years, its averages vary from 402 to almost 630 in. (10,200 mm to almost 16,000 mm).

MOST RAIN FROM MINUTES TO YEARS

TIME	PLACE	YEAR	in. (mm)
1 minute	Sainte-Anne, Guadeloupe, Caribbean	1970	1.5 (38)
1 hour	Holt, Missouri, USA	1947	12.0 (305)
1 day (24 hrs.)	Cilaos, Réunion, Indian Ocean	1966	71.9 (1,825)
1 month	Cherrapunji, India	1861	366 (9,296)
1 year	Cherrapunji, India	1861	1,042 (26,467)

DRIEST CITIES

Arica in Chile (below), on the edge of the Atacama Desert, is probably the most arid city on the planet. Its average rainfall is less than 0.08 in. (2 mm) each year. Other contenders, all with fewer than 0.5 in. (5 mm) of rain annually, are Ica in Peru, also bordering the Atacama, Luxor, and Aswan in southeast Egypt, and Al Jawf, southeast Libya—all of which are located in Africa's Sahara Desert.

MOST ARID PLACES

There are two areas on Earth where some parts have seen no rain (or other precipitation) since weather measurements began. One is the Atacama Desert in northern Chile, South America. The parts of the desert between the Chilean Coastal Range mountains and the Andes Mountains (below) are in a double rain shadow (see pages 34, 56). The other most arid place is the McMurdo Dry Valleys in Antarctica. This area also lies in a rain (or rather snow) shadow. Fierce winds race along the valleys, blowing snow away so no glaciers can form. Instead, the valleys have a bare, gravel-covered surface in contrast to the snow and ice all around.

RAINFALLS COMPARED

Here's a list of average rainfalls for a selection of places from across the world, from wettest to driest. Some places, such as London in the UK, have a reputation for being very rainy, but the scientific measurements don't always agree.

LOCATION	YEARLY AVERAGE, in. (mm)
Mawsynram, India	508 (12,900)
Quibdó, Chile	320 (8,130)
Monrovia, Liberia	182 (4,624)
Hong Kong, China	94.5 (2,400)
Abuja, Nigeria	58 (1,470)
Rio de Janeiro, Brazil	50 (1,250)
Islamabad, Pakistan	40 (1,000)
Chicago, USA	35 (890)
World average (over land)	28.1 (715)
London, UK	23.7 (601.7)
Melbourne, Australia	21 (530)
Deserts (official definition)	10 (250)
Arica, Chile	Less than 0.08 (2)

CASE STUDY:

STRANGE RAINS

There are many old phrases, sayings, and stories about strange rainfalls. Accounts describe blood, sand, and small creatures such as worms, fish, frogs, birds, jellyfish, snake, and even mice and rats falling from the sky in heavy downpours. Many of these tales date from long ago and have probably been exaggerated in retellings. But there are plausible, scientific reasons for most of these bizarre events—although maybe they're not so mysterious or entertaining.

UP AND DOWN

One weather phenomenon that can explain some strange rains is the waterspout. Similar to a tornado (see page 106), it is a fast-swirling column, or funnel, of air and moisture that forms over water, usually under a thundercloud (right). A waterspout can be powerful enough to suck up water and small items very near the surface, which could explain why many animal rains are aquatic creatures. As the waterspout moves along over land, it loses its power and drops its contents.

STAINED RAIN

Colored rains are not uncommon. In southern Europe, rain carrying red, orange, or yellow dust falls from clouds that formed far to the south during powerful sandstorms in the Sahara of North Africa. These are then blown north and west by strong winds. Some "blood rain" is due to tiny, light, reddish spores of algae (simple plants), which live on rocks and trees. Released into the air in their millions, the spores are carried aloft by rising air currents and mix with rain.

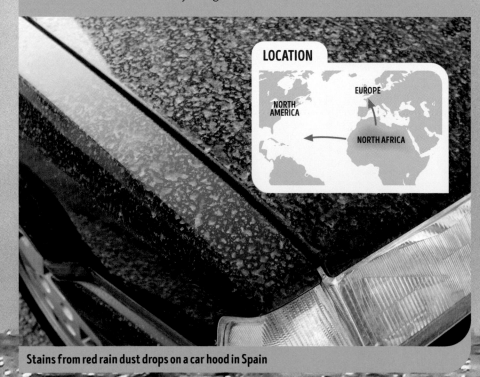

LOCATION

EUROPE

NORTH
AMERICA

NORTH AFRICA

Stains from red rain dust drops on a car hood in Spain

AGE-OLD TALES

Stories of strange rains go back almost as far as recorded history. People in ancient Egypt, China, and Central America all described these happenings. Many were woven into local customs and folklore, often involving gods and spirits. Sometimes, these have persisted and been updated into modern forms. Other descriptions of bizarre rains may be based on the mishearing of words. For instance, the origin of the English phrase "raining cats and dogs" may come from the Old English word *catadupe*, meaning "waterfall."

An early-19th-century cartoon, depicting the phrase "raining cats and dogs"

FISHY RAIN? MAYBE NOT

Some "fish rains" may be due to waterspouts, but others can be explained by animal behavior. For example, in a dry period, fish or frogs can be forced into crowded pools as the water dries up. If it suddenly rains or floods, the animals' instincts are to swim away and disperse to new places. When the heat returns and the waters dry up again, the animals are left stranded as though they fell with the rain, like these fish stranded after a flash flood.

UP NOT DOWN

Fish rains occur regularly in some places—but the fish do not come down, they go up! Heavy seasonal rains swell underground pools and rivers. Their waters flow up channels onto the surface, carrying cave-dwelling fish. The waters drain away and leave the fish exposed.

THE BIGGEST WINDS

Our planet's most powerful winds are short-lived tornadoes and hurricanes (see pages 106, 109), but some places are naturally windy for most of the time. These areas are usually over wide oceans or along coasts. Here, the flat surface presents little obstacle to air movement—unlike hills, valleys, trees, and other objects on land that act as windbreaks to slow moving air. Hilltops and mountain peaks are also very windy, since wind speeds increase with altitude (see page 15).

ROARING TO SCREAMING

Only one-third of Earth's land is located south of the Equator, in the Southern Hemisphere. This means that winds there have more flat ocean to blow across than in the Northern Hemisphere, so they are generally faster and more sustained. During the "Age of Sail," from the 16th to the mid-19th centuries, speedy sailing ships like *Cutty Sark* (shown here) raced to carry cargoes worldwide. The sailors gave these southern winds various names (see below).

➡ ROARING FORTIES

Westerly winds, usually at latitudes 40–50° south.
These are due to warm air masses rising at the Equator and cooler air moving in to replace them, known as the northern, or midlatitude, cell—which is driven like a cogwheel by the Hadley and polar cells on either side (see page 18).

➡ FURIOUS FIFTIES

Similar winds to the Roaring Forties, but at higher latitudes, 50–60° south.
These are due to the southern midlatitude cell (see page 18), and are almost uninterrupted all around the planet.

➡ SCREAMING SIXTIES

Westerly winds, like those above, but often even stronger.
These are at latitudes 60–70° south, where the midlatitude cell meets the Antarctic polar cell.

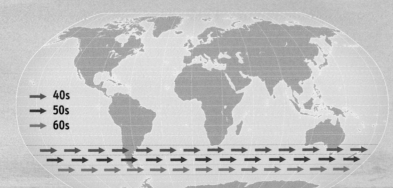

➡ 40s
➡ 50s
➡ 60s

WINDY CITY: 1

Chicago (below), on the shore of Lake Michigan, has long been called the Windy City. There are various stories about how it came by its name. One is that Chicago's politicians and business leaders were "windy" people—as in braggers and boasters. Another is that during hot summers, a pleasant cooling breeze wafts across the open expanse of the lake. However, the city's yearly average wind speed is just 10.2 mph (16.5 km/h), which is only slightly more than Central Park in New York.

WINDY CITY: 2

Pamplona in northeast Spain has a reputation as one of Europe's windiest cities. It has a relatively high altitude of 1,480 ft. (450 m), and receives winds from the Atlantic Ocean that funnel and strengthen along the Basin of Pamplona valley. In winter months, speeds average 16 mph (25 km/h), and on some days can be twice that. Wind turbines have been installed along the basin's high ridges (above).

WINDY CITY: 3

Wellington, capital of New Zealand, is at the southern end of North Island, on the Cook Strait, the narrow seaway separating the country's North and South Islands. At latitude 41° south, it is in the Roaring Forties. Winds from South America accelerate across the vast southern oceans and are then funneled and intensified through the "wind corridor" of Cook Strait. This makes Wellington one of the world's windiest cities, with hilltop speeds in some years averaging nearly 20 mph (30 km/h).

The *Solace of the Wind* statue depicts a person leaning into the gusts at Wellington waterfront.

WHAT IS A KATAB?

The greatest sustained wind speeds in Antarctica blow faster than 75 mph (120 km/h) for weeks. Coastal areas experience katabatic (meaning "going down" in Greek) winds. These are caused by air in the central, coldest parts of Antarctica cooling so drastically that it becomes dense and heavy. It then slides down from the high ice sheets of the center to the sea, gaining speed on the way. Some of these bone-chilling blasts can reach 185 mph (300 km/h).

People build a snow wall to protect a tent from high winds near McMurdo Station, Antarctica.

DROUGHT DESTRUCTION

Life on Earth depends on liquid water. So a lack of the usual rainfall can be one of the greatest of all weather-related catastrophes. It affects not only ourselves and our farm livestock and food crops, but also all wildlife. Like a flood (see page 98), a drought is a comparative, or relative, extreme weather event. This means that it occurs when the regular local weather conditions—in this case, the amount of rain, snow, or other precipitation—is much changed for a prolonged period of months or years.

Dry season

Rainy season

Scar from
forest fire

First-year growth

DROUGHT AND TREES

Large, healthy trees can usually survive several years of below-average rainfall. But their growth is slowed by this "water stress." The tree's rings inside the trunk record such events. During each yearly growing period, warm, moist conditions form a wider ring. A dry and either unusually cool or exceptionally hot year produces a narrow ring. Several years of drought make a cluster of narrow rings.

[◎] Buried machinery at a farm in South Dakota, in 1936, during the Dust Bowl disaster

DROUGHT AND DUST

During the 1930s in North America's Great Plains, the Dust Bowl disaster caused massive ecological, social, and financial ruin. Several years of low rainfall, combined with farming methods that destroyed the soil's structure and nutrients, produced fine, dry dust that blew in the wind, so that nothing could grow. Today's experts believe the disaster was partly caused by several rapid, extended La Niña events in the Pacific Ocean (see page 24), combined with unusual warming due to the Atlantic Multidecadal Oscillation (see pages 24, 95).

DROUGHTS IN HISTORY

There have been many devastating droughts throughout human history. Here are some of the worst:

○ NORTH AMERICAN GREAT DROUGHT, 1276–1299

Native Americans abandoned their crops and settled communities, and adopted a nomadic lifestyle in search of food.

○ NORTHERN CHINESE FAMINE, 1875–1880

It's possible that 12 million people died as crops failed. India and Africa were also badly affected, with some 8 million lives lost in the Indian Great Famine. Causes included a powerful Pacific El Niño and a similar cycle across the Indian Ocean, called the Indian Ocean Dipole.

○ FEDERATION DROUGHT, 1895–1903

Disturbed El Niño/La Niña cycles brought years of increasingly low rainfall across the eastern half of Australia. It devastated farm animals and crops, with almost no wheat harvest at all in 1902.

○ POVOLZHYE DROUGHT, 1921–1923

Along Russia's Volga and Ural Rivers, due partly to unusual jet streams, about 5 million people died, already weakened by the effects of World War I and the Russian Revolution.

Starving Russian families in the Volga area during the Povolzhye Drought

DROUGHT AND THE HUMAN COST

In eastern Africa in the early 1980s, Ethiopia endured three years of terrible famine in which more than 1 million people lost their lives. This drought-prone region has experienced many similar periods before and since, due to weather influenced by global events such as El Niño/La Niña and the Indian Ocean Dipole (see left). However, the 1983–1985 disaster was greatly worsened by the military government's policies and the repeated civil uprisings they caused.

SAHEL AND PETORCA: CONTRASTING FORTUNES

Certain places have naturally low rainfall. Unless suitable drought-resistant crops are planted, farming is a gamble. In Africa's Sahel region, this danger is never far away. However, recent eco-measures called "greening the Sahel" may be helping to change the situation. In contrast, Chile has faced a mega-drought since 2010. In central Chile's Petorca Province, newly introduced crops, along with new and much-criticized systems of land ownership and water management, have greatly worsened the predicament.

LOCATION

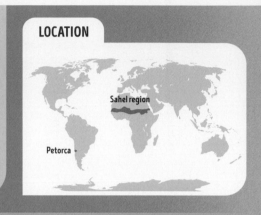

Sahel region

Petorca

DESERTIFICATION

Africa's Sahel region (main picture) is a landscape of dry grasses, scrub, and bush. It is a transition between the vast Sahara Desert to the north and the woods and tropical rainforest to the south. Above-average rains in the 1960s tempted people to farm the northern Sahel. But by 1970, the rains had returned to normal—only 6 in. (150 mm) a year in places. Continued growing of unsuitable crops destroyed the soil and vegetation, turning it into a desert-like wasteland, a process known as desertification. This led to famines in the 1970s.

SAHARA DESERT

SAHEL REGION

AFRICA

GOING GREEN

With international guidance and aid, from the 1980s onward, the Sahel's farmers began to use more suitable low-irrigation crops and to better manage soils and water resources. Water vapor given off by the plants rises, cools, and condenses into clouds, reducing the effects of the Sun's scorching rays and surface water evaporation. But natural climate phases may still cause problems. Some experts predict effects from what they call the Atlantic Multidecadal Oscillation (AMO). This is an El Niño-like cycle in the Atlantic Ocean that has a time period of between 60 and 80 years. However, other climate scientists believe the AMO is too weak and erratic to have much effect so far inland.

2.5 ACRES OF AVOCADO TREES USE AS MUCH WATER DAILY AS **1,000 PEOPLE**.

FALLING FAST

In 2020, some weather stations in the central provinces of Chile recorded only 1.97 in. (50 mm) of rain, compared to their average annual rainfall before 2010 of 9.84 in. (250 mm). Experts predict that climate change due to human activity could well continue this downward trend.

STOLEN WATER

Recently, Chile has suffered years of increasing drought, partly blamed on climate change. However, there are other factors involved. Around Petorca, avocado plantations have become common on the hillsides. The trees can only grow with plenty of water. There are accusations that rich landowners who set up and run the avocado groves persuade water companies to supply their hillside crops first, leaving less water for ordinary people in the lowland areas around.

WILDFIRE DISASTER

One of the most worrying predictions of global warming and climate change is an increase in extreme weather events leading to wildfires. Prolonged drought, often with prolonged heat, dries the land and air. Then, intense "dry" (without rain) thunderstorms bring lightning strikes that ignite the parched vegetation. However, up to nine-tenths of wildfires do not begin naturally and are due to human activity. With forecasts for more extreme weather, and a growing population, wildfire destruction looks set to increase around the world (see page 144, 168).

WHAT WILDFIRES NEED

Conditions that increase wildfire risk include:

- Reduced or no rain for weeks or months. The soil dries and vegetation, such as dead leaves, stems, twigs, and bark, shrivel to become a "litter bed" and fuel for the fire.
- Increased temperatures, which often accompany drought since there is less rain cloud cover to deflect the Sun's heat.
- Low air humidity, again common in a drought.
- High winds that fan the flames and spread them to new areas of vegetation.
- Some kind of ignition or spark.

FANNING THE FLAMES

Wildfires include bushfires, brushfires, forest fires, and grass fires. Nearly all begin with a small spark—which can be caused by a metal train wheel on a railroad track. Wind plays a crucial role, too. Moving air brings the fresh oxygen that the fire requires to burn. The situation can become increasingly dangerous when wind fans the flames toward yet more dry, combustible material, rather than toward some form of firebreak, such as rocks or water.

A wall of flames advances in Australia's Blue Mountains during a bushfire in 2019.

NATURAL WILDFIRE IGNITERS

- Lightning strikes, especially on tall trees, are probably the most common ignition for natural wildfires. After the dry season, early thunderstorms often lack rain.

- Superheated lava (molten rock) flowing from an active volcano can set dry vegetation alight.

- Sparks and lumps of lava known as "volcanic bombs" thrown out by a volcano may start a fire several miles away.

- Spontaneous ignition of tinder-dry vegetation can be caused by marsh gas, or methane, drifting from a swampy area nearby.

- Shiny natural objects, such as minerals containing gems, sometimes reflect, or focus, the Sun's rays, starting fires.

- Intense heat from coal seams that have been smoldering belowground, perhaps for centuries, can reach a surface outcrop covered with dry vegetation and set it alight.

FIRE CONTROL

Firefighters may not have long to plan tackling a blaze. Water, foam, and beating the ground at the site—along with the creation of firebreaks ahead of the flames using bulldozers or controlled burns—are just a few methods they can deploy. They can also drop water and fire retardant powder from helicopters and aircraft, like this water-bomber in action over a wildfire in Canada.

CARELESS PEOPLE

Wildfires due to human activity are increasing as climate change brings more heat and drought, and also as more people travel to wilderness areas. Major causes include sparks from campfires or fires that have not been put out properly, as well as heat and sparks from vehicles and generators, firearms shooting, cigarettes, and broken power lines. Perhaps the most difficult to understand is arson. At least 1 in 10 bushfires in Australia is started deliberately, according to the nation's National Centre for Research in Bushfires and Arson.

Smoke from land clearances in Brazil's Amazon Rainforest, 2019

SPREADING FAST

Satellites track and take images of wildfires around the world. This can help identify deliberately set fires, such as areas of tropical forests that have been torched to clear land for farming, quarrying, and mining. In the USA in 2019, there were about 55,000 problem wildfires destroying around 4.7 million acres (1.9 million hectares). The next year, there were almost 59,000, destroying twice that area.

FLOOD CATASTROPHE

Rainfall varies hugely with location and season. In most areas, people and nature have adapted to its average amounts, and they can also cope with the occasional increase or reduction. But sometimes extreme weather brings far too much rain, resulting in a flood. Like a drought (see page 92), a flood is a comparative, or relative, extreme weather event. It occurs when there is much greater rainfall, or other precipitation, than usual over a short period of days or hours—or maybe even minutes.

NATURAL FLOOD CYCLE

In Botswana, Southern Africa, the Okavango River never reaches the sea. Instead, it flows into a wide, low inland area on the edge of the Kalahari Desert called the Okavango Delta. Each summer, the river carries heavy seasonal rains from the highlands to the north. The water floods over the delta to extend its swampy area to a vast 5,000–7,700 square miles (15,000–20,000 sq km). Over several months, the water evaporates and drains, and the delta shrinks until next year. Plants and animals have long adapted to this yearly cycle.

[◎] The flooded Okavango Delta in the Kalahari Desert, Botswana, Africa

NOT ONLY THE WEATHER

Waterways are dammed for various reasons, including storing water for people, farming, and industry, and for hydropower to generate electricity. Exceptional rainfall may accumulate so much water that it breaches a dam, flooding the area below. But extreme weather is only part of the problem. Poor design, ignoring legal requirements, faulty materials and workmanship, and bad maintenance have all led to dam failures. When the Patel Dam in Kenya burst after heavy rain in 2018, the resulting floods caused a huge amount of damage (left), destroying villages and a school, and killing 50 people. The cause of the disaster was traced back to faulty dam design—for instance, there were no overflow channels to allow excess water to escape.

STORM FLOODS

Tropical storms (cyclones and hurricanes, see page 108) can bring massive deluges in a short time and are a common cause of flooding, even in regions used to their arrival. In 2011, Tropical Storm Washi dumped more than 16 in. (400 mm) of rain in 24 hours on parts of the Philippines, causing huge damage (shown below in the town of Cagayan de Oro). Rivers were overwhelmed and flash floods spread, raising water levels by more than 16 ft. (5 m). In excess of 1,500 people died and more than 100,000 were made homeless. Experts estimated that Washi was a once-in-50-years event.

BLOCKED DRAINS KILL

In built-up areas, especially with huge areas of water-resistant asphalt and tarmac, it is essential to design drainage that can cope with exceptional rainfall. An increasing cause of floods is blocked storm drains, gullies, and channels that are supposed to guide away excess water. In 2021, a fierce hailstorm hit the city of Sucre in Bolivia (above). The storm drain system was blocked with trash and debris. The ensuing flash floods turned streets into rivers, killed several people, swept away vehicles, and ruined roads and houses.

SNOWED IN, SNOWED UP

Like droughts and floods, excessive snow is a relative, or comparative, extreme weather event. It occurs when a location has a lot more snowfall for a much longer period than usual (see pages 36–37). Typically, a snowfall is 10 times deeper than the same amount of rainfall, although this varies, depending on the type of snow—for example, whether it is fluffy or compacted. Serious snowfall can paralyze a whole country, halting travel and transportation, taking out power and energy systems, and bringing shortages of food and water—as well as deadly freezing conditions.

TRAVEL TROUBLE

Few weather conditions disrupt travel as rapidly or as severely as snow and ice. Roads become impassable within minutes. If freezing temperatures persist, people and vehicles may be stranded for days. Weather forecasts carry warnings and often advise people not to make journeys, yet people still venture out. These motorists have been caught out by a sudden snowstorm on a mountain road in southwest Poland.

SNOWIEST CITY

The coastal city of Aomori in northern Japan (above) is often rated as the world's snowiest city. Its latitude of 40°N is similar to Chicago in the USA, Madrid in Spain, and Beijing in China. Yet it receives much more snowfall—over 300 in. (760 cm) each year—and it's all down to geography. In winter, extreme winds race in from Russia to the northwest, while in the south, cold air from the mountains flows northward. These two fronts combine, making the air rise and cool, and causing snow to form (see page 33).

PLENTIFUL WEATHER

The most eastern mainland city in North America is St. John's, in the Canadian province of Newfoundland and Labrador (below). It is also Canada's oldest city—and one of its cloudiest, foggiest, and windiest! Annual snowfall exceeds 126 in. (320 cm). Most of it comes from storms heading northeast along the coast. Famously, almost every kind of precipitation can fall in St. John's in an hour: rain, snow, sleet, freezing rain that coats the ground with ice... then snow again. In January 2020, a record snowfall of 30 in. (75 cm) in one day caused a week-long state of emergency.

DEADLIEST BLIZZARD

One of the most deadly blizzards in recorded history occurred in south Iran in February 1972 (right). More than 26 ft. (8 m) of snow fell during the winter storm and remained for a week, killing an estimated 4,000-plus people. Several towns and villages were totally buried by the unexpectedly massive snowfall, leaving no survivors.

Snow is common in parts of Iran but the amounts in 1972 were unprecedented .

DEEPEST DRIFTS

Winds can pile up loose, fluffy snow into huge drifts and mounds. Several locations claim the deepest-ever drift. In 1911, a huge bank of snow in Tamarack, California, was measured at 37.6 ft. (11.4 m). Sixteen years later, in 1927, Mount Ibuki in Japan recorded a drift of 38.8 ft. (11.8 m). In more recent years, a tourist trail has grown up through Yuki-no-Otani in Tateyama, central Japan (below). Here, at an altitude of 8,200 ft. (2,500 m), a road is cleared for vehicles through snow mounds that can exceed 59 ft. (20 m) in height and stretch for more than 0.6 miles (1 km).

THUNDER AND LIGHTNING

Flash, crack, BOOM! There are few more breathtaking displays of nature's power than a thunderstorm. Driven by very warm, fast-rising air, it often brings a sudden, noisy, soaking end to hot, dry weather. A typical storm is more than 15 miles (25 km) wide and can have dozens of lightning flashes during its few hours of life. But a really huge thunderstorm is 10 times bigger and can last two or three days. At any moment, more than 2,000 thunderstorms are raging all around the world.

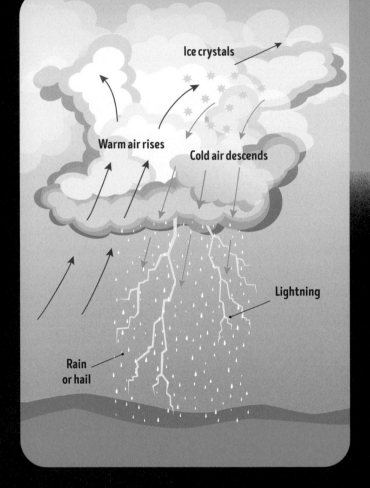

Ice crystals

Warm air rises

Cold air descends

Lightning

Rain or hail

INSIDE A THUNDERCLOUD

Thunderstorm, or cumulonimbus, clouds (see page 28) are huge and tall. Heat from the ground makes air rise fast and cool quickly in the upper atmosphere. As the cloud's water droplets and ice crystals rub past one another, they generate a negative electrical charge, which builds up and suddenly jumps out of the cloud as a giant spark of lightning.

TYPES OF LIGHTNING

CLOUD TO GROUND: The electric charge zigzags to the ground to release its electricity, then returns as the main flash.

CLOUD TO CLOUD: Different areas in nearby clouds exchange their electricity, which shows from a distance as "sheet" lightning.

CLOUD TO AIR: The electric charge is so strong it has to leap somewhere—usually sideways from the main cloud into midair.

SPRITES, JETS, AND BALLS

Before space rockets and satellites were invented, no one knew what happened above a thundercloud. We now know there are many more kinds of lightning-type electrical discharges in a range of colors, which take place in the upper atmosphere.

Red sprites are orange-red flashes high above a thunderstorm, sometimes going all the way into space.

Blue jets are another kind of upper atmospheric lightning that occurs in the stratosphere.

Ball lightning is a glowing shape that can move, even at ground level. But how it happens is still a mystery.

AWESOME SPARKS

Lightning can reach 250 million volts—about a million times more than household electricity. The flash has a temperature of more than 50,000°F (30,000°C). It heats the air around it, which expands faster than the speed of sound, creating a sonic boom we hear as thunder.

LIGHTNING STRIKE

Although it looks like we see lightning striking the ground from clouds, what we actually see is a powerful electrical current running the other way: from the ground to the cloud. First, an invisible channel of negative charge will make its way from the cloud to the ground in a millisecond. When it connects with a positive charge on the ground—such as on a high building—that's what creates the bright return flash that we see.

THE GODS ARE ANGRY!

Before we understood that lightning is electricity, many cultures believed that it was created by gods, perhaps to punish the sins of ordinary people. In the Norse mythology of Scandinavia, Thor was the god of thunder, lightning, and storms.

Cloud-to-ground strikes

Lightning strikes over Catatumbo, Venezuela. There is more lightning here than anywhere else on Earth, with around 150 storms a year.

The Norse god Thor depicted in a 19th-century painting

WEIRD WEATHER

Some of the most bizarre sights in our sky and across the landscape are actually due to weather events. Throughout history, these sights have been mistaken for gods, warring giants, ghostly spirits from other worlds, and even alien spaceships. In ancient times, these happenings were sources of wonder, fear, and worship. Today, modern science can explain most of them.

MIRAGES

A mirage is a trick of the light that usually makes objects in the sky, or just above the ground, appear to be on the horizon or on the ground in front of the viewer. It occurs because light rays from the object are bent and bounced by layers of hot air shimmering with heat near the ground. The object appears upside down, blurry and shaky, and looks lower than its true position. A cloud or even the sky itself may appear like a pool of water on the ground.

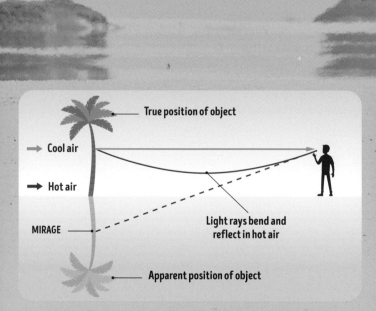

True position of object

Cool air

Hot air

MIRAGE

Light rays bend and reflect in hot air

Apparent position of object

MIRAGES IN MYTHS AND LEGENDS

Various myths and mysteries are believed to have their origins in mirages. For example:

- ☀ The *Flying Dutchman*, a terrifying ghost ship doomed to sail the seas forever.

- ☀ Some sightings of UFOs (unidentified flying objects) or UAPs (unidentified aerial phenomena).

- ☀ Fata Morgana—a complicated type of mirage where objects appear to hover above the horizon. These were once said to be the work of the sorceress Morgan Le Fay, from the stories of King Arthur. Fata Morganas may also have given rise to the popular idea in fairy tales of "castles in the air."

WHAT IS A RAINBOW?

Long believed to lead to a pot of gold, rainbows occur because of the way sunlight passes through raindrops. Inside each drop, the light rays are dispersed, or split, into their various colors (wavelengths), then bounced back the way they came. The path each different colored ray takes means that we see only one color at each vertical angle of our view—red highest, then orange, yellow, green, blue, indigo, and violet lowest, on the inside of the curve. To see a rainbow, the observer needs to have the Sun behind them.

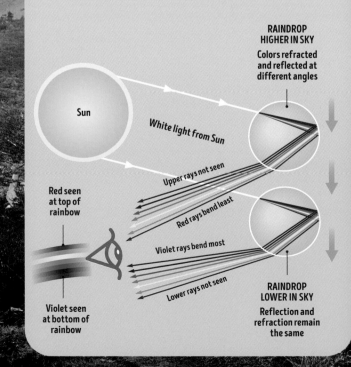

RAINDROP HIGHER IN SKY

Colors refracted and reflected at different angles

Sun

White light from Sun

Upper rays not seen

Red seen at top of rainbow

Red rays bend least

Violet rays bend most

Lower rays not seen

Violet seen at bottom of rainbow

RAINDROP LOWER IN SKY

Reflection and refraction remain the same

SUNDOGS

A sundog is a "look-alike sun" that appears next to the real Sun when it is low in the sky. Often, sundogs form in pairs to the left and right of the Sun. They are caused by sunlight being refracted, or bent, by ice crystals floating in some kinds of clouds, such as cirrus clouds. The crystals need to be in a certain position, with their large, flat faces parallel to the horizon. In some conditions the light forms a halo, or ring, around the Sun (right). Sundogs and haloes are known as parhelia.

BROCKEN SPECTERS AND GLORIES

Walkers in hilly places with both sunshine and clouds are sometimes startled to see a huge figure in the distance, perhaps with a rainbow around their head, like a halo (left). This strange apparition is actually a shadow of the person, cast by the Sun behind them, onto the cloud, mist, or fog in front. The name Brocken Specter comes from Brocken Mountain in Germany where this illusion is common. The rainbow halo is known as a glory. The same effect can happen with any object, including aircraft casting shadows on the clouds below.

SPINNING AND TWISTING

There are several kinds of extreme windstorms that rush across the Earth's surface. They vary in size from swirling dust devils barely bigger than a human being to fearsome tornadoes, or twisters, that can wreck houses and powerful derechos many miles long (see next page). Dust devils and tornadoes are types of whirlwinds with spinning air currents, known as vortexes. In contrast, the winds of a derecho blow mainly in one direction. The planet's speediest winds are tornadoes, which can whirl around at more than 300 mph (480 km/h).

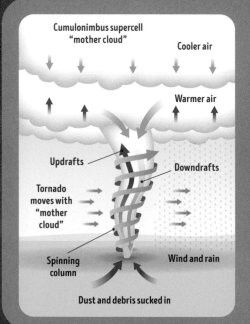

Cumulonimbus supercell "mother cloud"

Cooler air

Warmer air

Updrafts

Downdrafts

Tornado moves with "mother cloud"

Spinning column

Wind and rain

Dust and debris sucked in

THE DREADED TWISTER

A tornado is a fast-spinning column of air, often with a wide funnel-like top (left), that connects to a cumulonimbus "mother cloud" (see page 102). Warm, moist air rises quickly and encounters winds that change a lot with height, making the air spin. This spinning is increased by cold descending air. A tornado has formed when this spinning becomes a narrow, stretched column that reaches the ground. Cool air sinking from the mother cloud forms into a funnel that links up with the rotating warm air. As the funnel cloud reaches the ground, a tornado is "born."

TRI-STATE DEVASTATION

The Tri-State Tornado of 1925, affecting Missouri, Illinois, and Indiana, holds several infamous records:

🌪 Longest known path for one tornado, at 217 miles (350 km).

🌪 Longest duration, at about 3.5 hours.

🌪 Fastest speed moving forward, at 75 mph (120 km/h).

🌪 Most recorded deaths in US history from one tornado: 695.

AVERAGE TORNADO WIDTH:	330–660 ft. (100–200 m)
AVERAGE TORNADO FORWARD SPEED:	30 mph (50 km/h)
AVERAGE ROTATING WIND SPEEDS:	70–155 mph (110–250 km/h)
AVERAGE TORNADO PATH LENGTH:	less than 6 miles (10 km)

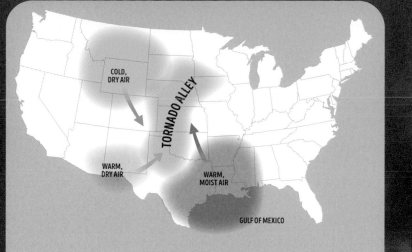

TORNADO ALLEY

Central North America experiences many tornadoes, especially during spring and summer, in an area known as Tornado Alley. In some years, the region can experience more than 1,000 of them. The tornadoes are caused when warm, moist air heading north from the Gulf of Mexico meets cold, dry air moving south from northern parts of the continent, and the flat landscape encourages their formation.

WHIRLING "DEVILS"

Dust devils (below) are small, narrow, swirling air columns, rarely more than 165 ft. (50 m) high. A small pocket of extra-warm air at the surface—for example, above rocks or blacktop heated by the Sun—rises. Passing winds start it spinning. More warm air sucks in at the base, picking up loose bits of dust or debris. Usually the source of extra-warm air soon fades, so most dust devils last only a few minutes.

DERECHOS

From the Spanish for "right" or "straight," derechos are powerful winds that blow in a straight line in front of fast-moving thunderstorms. They form when sinking cool air in front of thunderclouds burrows under warm air near the ground. The rising warm air sets up strong updrafts which, when cooled, drop back down to the surface to produce powerful downdrafts. A typical derecho is 250 miles (400 km) long, with gusts exceeding 62 mph (100 km/h) along most of this length, although speeds of 125 mph (200 km/h) have been known. They can cause as much damage as tornadoes.

Thunderclouds threaten to develop into a derecho over a beach in Florida.

 A tornado forms in a field in Colorado.

THE BIGGEST STORMS

It is difficult to comprehend the utter devastation that can be caused by our planet's most extreme storms—unless you experience one. Hurricanes and tropical cyclones can be the size of entire continents. They bring torrential rain, unstoppable winds, widespread floods, gigantic coastal waves and storm surges, and damage and death on a vast scale.

Eye of the storm

NAMES FOR TROPICAL CYCLONES

Tropical cyclones have a far lower atmospheric pressure compared to cyclones in temperate zones. They form all around the Tropics, but the storms they cause have different names in different regions:

⟃ HURRICANE

Probably from the Native American Taino word *hurakán*, meaning "god of the storm" or "evil wind spirit."

⟃ TYPHOON

Mixed origin, from the Arabic word *ṭūfā*, meaning "whirlwind," and the Chinese term *tai fung*, meaning "big wind."

⟃ CYCLONE

From the ancient Greek word *kuklos*, meaning "moving in a circle."

CHURNING FROM SPACE

Satellite images show the enormous scale of tropical cyclones. In 1975, Supertyphoon Nina, shown here, covered much of the western Pacific Ocean. As with most of these weather systems, the central area, or eye, is relatively clear, with light air movement. Around it swirl the fierce churning winds, clouds, and downpours that cause the greatest damage. One of the worst cyclones on record, Nina lasted nine days, with winds peaking at 155 mph (250 km/h). Its torrential rainstorms triggered dam collapses in China and caused huge amounts of damage across East and Southeast Asia. More than 220,000 people died.

HOW HURRICANES AND CYCLONES OCCUR

As the Sun heats the warm tropical ocean, typically to a surface temperature of 80°F (27°C) or more, the warm, moist air above it rises. As it does so, the water vapor in it condenses to form clouds and rain. The area of low pressure below the clouds sucks in more air, which is also heated and rises, and so it continues. Initially, light winds allow the storm clouds to rise to enormous heights, while the Coriolis effect (see right) starts the system spinning. Continued rising hot, humid air feeds the system until it becomes a tropical storm, then a full-scale tropical cyclone.

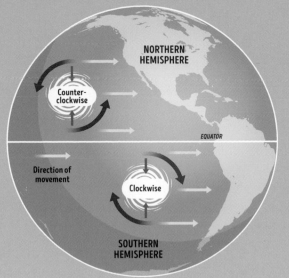

NORTHERN HEMISPHERE

Counter-clockwise

EQUATOR

Direction of movement

Clockwise

SOUTHERN HEMISPHERE

WHY STORMS SPIN

The ball shape of planet Earth and the fact that it rotates from west to east (left to right) mean that its surface moves fastest at the Equator and more slowly toward each pole (see page 48). This has a bending, or deflecting, effect on winds. Air initially blows from high to low pressure, but in the Northern Hemisphere air is deflected to the right so that it spins counterclockwise around a cyclone. The opposite happens in the Southern Hemisphere, so that cyclones spin in a clockwise direction.

HOW A HURRICANE FORMS

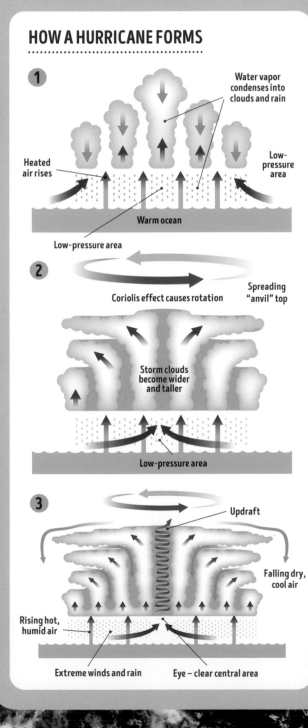

1

Water vapor condenses into clouds and rain

Heated air rises

Low-pressure area

Warm ocean

Low-pressure area

2

Coriolis effect causes rotation

Spreading "anvil" top

Storm clouds become wider and taller

Low-pressure area

3

Updraft

Falling dry, cool air

Rising hot, humid air

Extreme winds and rain

Eye – clear central area

TROPICAL CYCLONE STATS

There are no such things as "average" hurricanes or cyclones, but they often share some general features:

Height: The whole system may be 9 miles (15 km) tall

Width: Typically 310–370 miles (500–600 km)

Forward speed: About 9–12 mph (15–20 km/h), occasionally 40 mph (65 km/h)

Early, weak stages: Known as a tropical depression

Winds more than 39 mph (63 km/h): Known as a tropical storm

Winds over 74 mph (119 km/h): Known as a tropical cyclone

THE SAFFIR-SIMPSON SCALE

This scale ranks hurricanes and tropical cyclones according to their size, wind speed, and destructive power.

CATEGORY	WIND SPEEDS	
1	74–95 mph (119–153 km/h)	Some damage to roofs and weak structures, large tree branches torn off, power lines and poles damaged
2	96–110 mph (154–177 km/h)	Buildings damaged, shallow-rooted trees uprooted, power outages for days
3	111–129 mph (178–208 km/h)	Even well-built structures damaged, many trees toppled, electricity and water disrupted for days or weeks
4	130–156 mph (209–251 km/h)	Many buildings destroyed, most trees uprooted, power and other utilities may be off for weeks to months
5	157 mph (252 km/h) or more	Widespread destruction of all kinds

CASE STUDY:

KATRINA: TITANIC FLOODS

Hurricane Katrina began in the western mid-Atlantic as a tropical depression on August 23, 2005. The following week brought terrible developments. Katrina became a named tropical storm on August 24, and the day after, it was designated an official hurricane as it passed over Florida. Moving west, it was fueled by the Gulf of Mexico's abnormally warm waters. On August 28, it reached maximum Category 5 strength (see page 109). The next day it crossed the Louisiana coast, heading north to create havoc on a massive scale. Katrina gradually faded to storm strength about 155 miles (250 km) inland.

LOCATION

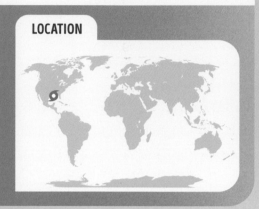

WATER EVERYWHERE

The low-lying city of New Orleans (right) was hit especially hard by Katrina. The levees (banks) designed to protect the city from the sea were broken by huge waves and tidal surges. More than three-quarters of the city and suburbs were under seawater. The damage was colossal. The floodwaters destroyed buildings and transit systems and remained for weeks, ruining vast areas of farmland. The disaster caused a total of 1,800-plus fatalities.

THE END OF THE WORLD

At full strength, approaching Louisiana, Katrina's winds reached 172 mph (275 km/h). Even when the storm reached the coast it was still a Category 3. It was also exceptionally large, covering an area of almost 250 miles (400 km) across with hurricane-speed winds. Storm conditions persisted until the weather system reached the Great Lakes of North America, almost 1,250 miles (2,000 km) to the northeast.

KATRINA FROM SPACE

Satellites tracked Katrina's size, path, and wind speeds for more than one week (right). It was the third main hurricane of the hurricane season, which is between June and November in the Atlantic, Caribbean, and Gulf of Mexico. The key below shows how the wind strength changed over the week, using the Saffir-Simpson scale (see page 109).

- TROPICAL DEPRESSION
- TROPICAL STORM
- CATEGORY 1
- CATEGORY 2
- CATEGORY 3
- CATEGORY 4
- CATEGORY 5

AUGUST 31

PATH OF HURRICANE

New Orleans

ATLANTIC OCEAN

FLORIDA

GULF OF MEXICO

AUGUST 23

NAMING TROPICAL CYCLONES

Cyclones are given names as well as codes to help people identify and discuss them more easily. In the Pacific and Indian Oceans, the names of animals, plants, and foods are used. In the Atlantic, most are given people's first names, in alphabetical order and alternating between female and male. In 2020, these began with Arthur, Bertha, Cristobal, Dolly, Edouard, and Fay. The names are chosen from lists already compiled by meteorologists and repeated every six years. However, the names of especially deadly storms, like Katrina, are "retired" and new ones chosen.

VOLCANOES AND WEATHER

In 1816, the world endured a "year without a summer." In the previous year, there had been a gigantic eruption of Mount Tambora volcano on the island of Sumbawa in what is now Indonesia. So much vapor and ash spewed into the atmosphere and drifted around the planet that it caused excessive clouds and haze to reduce the Sun's warmth the following year. Summers in some areas were up to 5.5°F (3°C) cooler. Other volcanic eruptions have affected the weather in similar ways, although perhaps not on such a massive scale.

BLOCKING THE SUN

Tambora discharged not only solid materials, from rocks as big as houses to tiny specks of dust and ash, but a suffocating cocktail of gases and vapors, too. These included sulfur dioxide, which reacted with oxygen in the atmosphere to create thick, fog-like chemicals that dimmed the Sun. Fine ash rose 19 miles (30 km) into the sky and wafted around the globe on jet streams and other winds. The death toll in the local area, including those from the resulting tsunamis (enormous waves), suffocation, hunger, and disease, probably exceeded 100,000. Crop failures and famines resulted in more fatalities as far away as Africa, Europe, and the Americas.

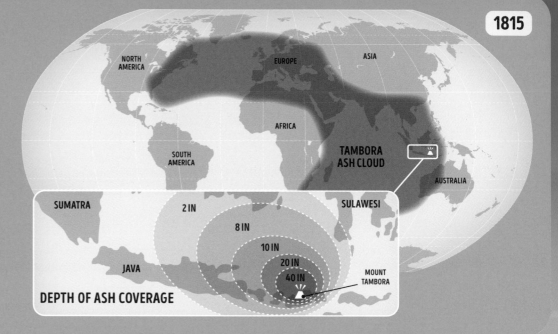

1815

NORTH AMERICA

EUROPE

ASIA

AFRICA

SOUTH AMERICA

TAMBORA ASH CLOUD

AUSTRALIA

SUMATRA

2 IN

8 IN

10 IN

20 IN

40 IN

JAVA

SULAWESI

MOUNT TAMBORA

DEPTH OF ASH COVERAGE

TAMBORA AFTERMATH

Tambora's eruption destroyed much of the volcano, leaving behind a giant crater, or caldera (shown here), about 4 miles (6 km) wide with a maximum depth of 3,600 ft. (1,100 m). Much of the island's plant and animal life was destroyed. However, after about five years, the weather in the area had returned to normal, plants were growing once more, and, within 15 years, people were able to farm again.

ASH AND MORE ASH

In June 1991, eruptions began from Mount Pinatubo in the Philippines (left), culminating in vast quantities of ash being ejected into the sky to heights of 25 miles (40 km). As the ash drifted and dispersed, it lessened the Sun's light and warmth, so that temperatures around the Northern Hemisphere reduced by 1.1°F (0.6°C) for many months. Because the ash changed the atmosphere's makeup, the ozone layer suffered faster thinning than was considered usual at that time.

BEAUTY AFTER DISASTER

Dust, ash, and other floating particles from volcanic eruptions have produced some amazing atmospheric sights. For a few years after Pinatubo, people were still astonished to see intensely colored sunrises, twilights, and sunsets. This was due to lingering volcanic particles in the atmosphere scattering and reflecting the sunlight. This photograph shows a colorful display in the atmosphere above the dormant volcano Mauna Kea on the island of Hawaii. It was caused by distant clouds from the Mount Pinatubo eruption casting shadows on dust in the upper atmosphere.

THROWN INTO THE AIR

- Tambora's eruption blasted more than 50 cubic miles (200 cubic km) of rocks, ash, and other particles into the atmosphere.

- This was about 10 times more than the 1883 eruption of Krakatoa, also in what is now Indonesia.

- It was also perhaps 15–20 times the amount ejected by Pinatubo in 1991.

- The eruption of Vesuvius in southern Italy in 79 CE, which buried the towns of Herculaneum and Pompeii, expelled an estimated 0.24 cubic miles (1 cubic km) of material.

USING THE WEATHER

From ancient windmills and waterwheels to the latest wind turbines and hydroelectric dams, people have long been exploiting the weather. Throughout history, a steady series of inventions and practices have sought to capture the weather's energy and other features for human use.

Some devices rely on the Sun directly—for example, solar panels, which utilize its light and heat. These are highly dependent on the weather, as they work best in sunny places. Other devices exploit the Sun's power indirectly. A dam's hydroelectric turbines harness the energy of a river's moving water. This water fell as rain in the water cycle (see pages 26-27), after being evaporated by the Sun's heat into water vapor. The warm vapor rose into the atmosphere, cooled, condensed, fell as rain, and flowed into the river.

The Itaipu hydroelectric dam on the Paraná River on the border between Brazil and Paraguay

WEATHER-POWERED ENERGY

Giant hydroelectric dams are a major feature of the landscape in some countries, including China, Canada, Brazil, the USA, and Norway. Hidden within are rows of massive turbines and generators that are spun around by the flowing water to produce the electricity. They are usually located inside a turbine hall at the foot of the dam, like the one on the right, which belongs to the Hoover Dam on the Colorado River, on the border between the states of Nevada and Arizona.

WEATHER-POWERED FUN

A long list of sports and leisure activities rely on moving water provided by the weather. One of the most exciting—and strenuous—is white-water rafting, shown here on the River Ganges in Rishikesh, northern India. As a white-knuckle recreation, white-water rafting began in the 1940s when suitable inflatable rafts, first developed for military service, became widely available.

WEATHER POWER IN HISTORY

Weather energy in the form of moving water and moving air has been exploited for at least 5,000 years. In North Africa, around 4,500 years ago, the River Nile was used to transport enormous blocks of stone from quarries to the building sites of the Great Pyramids—some of the granite came from Aswan, almost 625 miles (1,000 km) south. Around the same time, early sailing ships navigated the Mediterranean and also the Indo-Pacific region, leading to people discovering new lands and islands across South and Southeast Asia, and into the Pacific. From 2,000 years ago, inventions such as waterwheels and windmills took the physical labor out of tasks such as lifting water from wells and grinding grain. During the Age of Sail (see page 90), large sailing ships, mainly from Europe, reached new territories and established trade routes around the globe.

OLD-TIME WIND FARM

A typical windmill consists of large sails attached to the top of a tower. Wind spins the sails to drive machinery housed within the tower. These kinds of windmills have been used in Western Asia since at least the 8th century CE. They soon spread to Europe where, in the 12th century, a new type of mill was invented that could turn its sails in any direction to face the wind. Windmills were used for many purposes: to grind grain into flour, to raise water into channels for crop irrigation, to lift unwanted water over banks to flow away, to mash wood for papermaking, to treat fibers in textile production, and to work machines such as saws and lathes. These traditional 17th-century windmills are in La Mancha, Spain.

DUSTY MILLER

In the second millennium CE, windmills began popping up across Europe and Central and East Asia. By the 19th century, there may have been as many as 200,000 working windmills in Europe. The miller was an important person in the local community, using their mill to grind various products and power numerous machines. Today, there are only a few preserved working windmills left, such as this one in the Netherlands.

WIND-RAISED WATER

Many dry, remote areas have windpumps, like this one in South Australia, to raise underground water for people, livestock, and plants. Dating from about 1900, this multi-vane windmill works a mechanical pump that lifts water from deep belowground. It works day and night, while its rear vane turns to keep it facing the wind. The water is stored in tanks, and when they are full, a lever turns off the pump. It's a lot less tiring than operating a handle to lift a heavy bucket of water from a well!

PRESERVING FOOD

Farming created the need for food storage and preservation. Grains can be stored in jars, but other, moister foods, such as meat, fish, and vegetables, have to be dried first to reduce the risk of rot and decay. Traditionally, this is often done using the Sun and wind. Foods preserved in this way include sun-dried tomatoes, raisins (dried grapes), and biltong (dried meat from South Africa). These fish are being dried on racks on a beach in Nazaré, Portugal.

WATERWHEELS

Waterwheels of various designs have existed for more than 3,000 years. They had many purposes, including lifting water for irrigation and powering windmills. Waterwheels improved greatly in the 19th century when the mill race (shown above) was developed. This was a narrow channel that sent a fast stream of water to the wheel, helping it to turn more quickly. The pond provided water for people, animals, and crops, created a place to keep fish for food, and was also a reservoir so the waterwheel could continue working even during periods without rain.

SUNNY COOKING

The 1960s saw a trend for solar cookers. These consist of a shiny metal or plastic bowl that directs the Sun's heat onto a cooking pot. Of course, the cooker can only be used in sunny weather and needs protection from bad weather. This fold-out version from 1963, in use on Mount Everest, was called the Sundiner. Modern versions of solar cookers are helping to reduce energy use.

WIND POWER

In coming years, weather-related energy sources will be essential to meet the challenges of climate change (see pages 150–179). Groups of wind turbines, known as wind farms, will be a vital part of this process. They convert the kinetic energy (energy of motion) of flowing air into electricity. Thanks to weather satellites in space, and land- and sea-based weather stations around the world, wind-farm builders now have access to huge amounts of weather information. They can analyze the data to find the best wind-farm sites where there are suitably strong, regular, year-round winds, but a low chance of extreme weather.

OCEANS OF WIND

Strong, persistent surface winds blow across the flat open expanse of many seas and oceans. Sets of turbines, standing on the shallow seabed, are called offshore wind farms. Unlike onshore wind farms, they do not use up valuable land space or intrude on beautiful natural scenery. However, they are more costly to install, needing specialized ships and equipment for construction, and long undersea cables to carry the electricity ashore. The Borselle wind farm, shown here, lies off the coast of Zeeland in the Netherlands.

INSIDE A WIND TURBINE

Most wind turbines consist of a tall tower topped by a pod called a nacelle, which carries the main blades, or rotors. When the wind blows, these turn a low-speed shaft with gears attached. These increase the rotation rate onto a high-speed shaft to generate more electricity. A brake makes sure that strong winds do not rotate the shafts too fast, which would damage them. The direction of the wind is detected by an anemometer (see page 136), while electric motors turn the nacelle and rotors to face it. The blades change their angle, or pitch, to the wind, according to wind strength, to keep turning at the most effective speed. In high winds they can tilt almost edge-on to avoid damage.

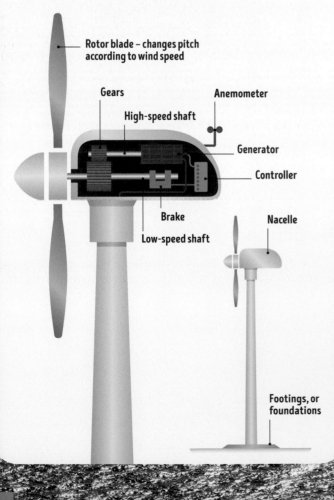

Rotor blade – changes pitch according to wind speed

Gears

High-speed shaft

Anemometer

Generator

Controller

Brake

Low-speed shaft

Nacelle

Footings, or foundations

Assembling the rotor blades of a new wind turbine in Normandy, France, 2014

THE GROWTH OF TURBINES

A typical medium-sized turbine generates around 1.5–2 megawatts of electricity—enough to power 500–1,000 homes in a developed country. The blades are generally 130–195 ft. (40–60 m) long, and mounted on a tower 245–280 ft. (75–85 m) tall. However, turbine sizes are gradually increasing. Some have blades exceeding 360 ft. (110 m) in length, and generate more than 15 megawatts—though weather damage to larger turbines is more costly to repair than for smaller turbines.

ONSHORE WIND FARMS

The first wind turbine to generate more than 1 megawatt of electricity was in Vermont in 1941. From the 1990s, onshore (land-based) wind farm growth has increased. China produces about one-third of all wind-generated electricity (both onshore and offshore), followed by the USA, Germany, then India. However, onshore wind farms in particular face criticism. Opponents say they spoil the view, produce noise pollution, are a danger to birds and bats, and reflections from the spinning blades can cause seizures in some people, especially when driving.

THE GROWTH OF WIND POWER

In recent decades, wind power has been the fastest-growing source of clean, green electricity. In some years its output has risen by more than 10 percent. In 2020, wind produced about one-fifteenth of the world's electricity. Among renewable sources of electricity, it was second only to hydropower (see page 120). But to meet future targets such as those set by 2021's COP26 Conference in Glasgow, Scotland (see page 171), wind and other renewables need to expand at two to three times their present rates.

An engineer sits on top of a wind turbine, surveying the landscape.

HYDROPOWER

Hydropower, or hydroelectricity, is electricity generated from moving water—and the biggest source of renewable energy. Like wind power, it uses the Sun's heat in an indirect way by harnessing the power of moving water, which has been set in motion by the Sun (see page 26). Unlike wind, hydropower can be captured in places that are far away from the weather that produced the original flow of water. This is usually at a barrier, or wall, across a river known as a hydroelectric dam, or hydrodam.

LOCATING A HYDRODAM

Hydroelectric dams must be carefully sited so that the flow of water is enough, both in quantity and strength, to make the generation of electricity worthwhile. They are usually built at a narrow part of a river's course, where the current is stronger and the dam can be narrower and therefore less costly to construct. Another major factor that also needs to be considered is the reservoir, or artificial lake of water, which forms behind the dam. This can provide year-round water for local farming, irrigation projects, and human use. But some hydrodams have flooded natural areas, farmland, villages, and even whole towns. People have been forced to move away—often against their will, and sometimes without proper compensation. Dams can also prevent wildlife from getting up and down the river.

HYBRID SOLAR-HYDRO

Hybrid solar-hydro power makes direct use of the Sun's heat. A solar farm of floating solar panels is installed on the reservoir behind the dam. These convert sunlight energy straight into electricity. Since 1971, the Sirindhorn Hydro Dam across the Lam Dom Noi River in northeast Thailand (below) has been producing 36 MW (megawatts, see page 123) of electricity. In 2021, more than 140,000 solar panels were installed, covering about 346 acres (140 hectares) of its reservoir. The panels more than double the dam's power output to 81 MW.

📷 An aerial view of the Solina Dam, Poland's largest hydroelectric dam, located on the San River. The lake behind is an important tourist attraction for the region.

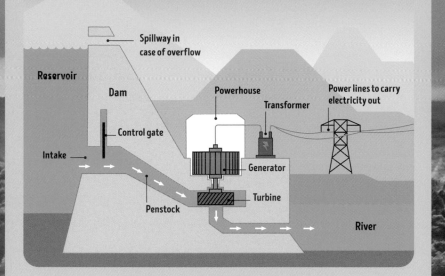

Spillway in case of overflow

Reservoir

Dam

Control gate

Intake

Penstock

Powerhouse

Transformer

Power lines to carry electricity out

Generator

Turbine

River

INSIDE A HYDROELECTRIC DAM

Far below the reservoir surface, water flows into the dam's penstock (channel) and is guided past the angled vanes, or blades, of a turbine, making them rotate. The turbine is connected to an electricity generator. Most hydroelectric powerhouses have sets of turbines. For example, the Hoover Dam across the Colorado River in the USA has 17 turbines and generators. Spillways allow excess water to drain away in times of flood, while the control gate can stop or restrict the flow of water.

DAM IN THE DESERT

The Aswan High Dam across the River Nile in Egypt (shown below in a satellite image) is in a desert. The rainy regions feeding the river are far to the south, some more than 2,000 miles (3,200 km) away in Rwanda, Burundi, and Tanzania. Between June and September, the dam's reservoir stores the rush of water that used to flood the Nile Valley and Delta. This water is gradually released throughout the year, giving much improved and controlled irrigation downstream for crops, livestock, people, and industry. The dam is 13,000 ft. (4,000 m) long, 360 ft. (110 m) high, and generates 2,100 MW of power.

Aswan High Dam

RENEWABLE POWER

⚡ Hydropower is a vital part of the challenge to tackle global warming and climate change.

⚡ It is the leading source of clean, green, renewable energy, generating almost one-fifth of the world's electricity.

⚡ The biggest producer is China, followed by Brazil, the USA, Canada, and India.

⚡ Ninety-nine percent of Norway's electricity comes from hydropower.

⚡ In theory, harnessing all the hydropower around the world that could be generated in a practical way would increase its use by 400 percent, significantly reducing the use of nonrenewable energy sources.

CASE STUDY:

THREE GORGES: POWER FOR MILLIONS

The world's biggest hydroelectric project is the Three Gorges Dam across the Yangtze River in southeast China. This gigantic system relies on the Yangtze's massive drainage basin that stretches west 700 miles (1,100 km) from the dam and reaches altitudes of more than 16,400 ft. (5,000 m). The basin includes the Tibetan mountains in western China where there are seasonal melting snows and glaciers, plus a monsoon climate in the southwest (see page 68), which has heavy rains from June to September. Formerly, these flooded the Yangtze Plain, causing vast damage and thousands of deaths. The dam now controls water flow throughout the year. However, its construction is still very controversial, as it destroyed important wild areas and submerged entire towns. It has also altered the river's silt and sediment levels, and increased risks of landslides and even earthquakes.

LOCATION

THREE GORGES SCENERY

The Three Gorges is a series of three narrow, steep-sided gulleys upstream of the dam. They cover an overall length of about 190 miles (300 km). Qutang Gorge (left) is farthest upstream and is 5 miles (8 km) long, Wu Gorge is 28 miles (45 km) long, and Xiling Gorge is 40 miles (65 km) long. The dam is sited in approximately the middle of Xiling Gorge. The giant dam and the surrounding spectacular scenery have made the area hugely popular with tourists.

CITY-SIZED DAM

The Three Gorges Dam (below) is 7,875 ft. (2,400 m) long and 1,590 ft. (80 m) high. Its reservoir behind covers more than 385 square miles (1,000 sq km) and stretches upstream around 370 miles (600 km). The dam took almost 30 years to construct, and was generating 22,500 MW of power by 2012.

📷 An upriver view of the Three Gorges Dam on the Yangtze River, China, showing the red flood control door lifts on top

DAM BYPASS

The Yangtze is the world's third-longest river and a busy working waterway. Massive ships travel between its mouth at Shanghai on the east coast, 1,000 miles (1,600 km) downstream from the dam, to the cities of west and southwest China. Two sets of staircase, or step, locks, each with a series of five 920 ft. (280 m) long lock zones (below), allow ships up to 11,000 tons (10,000 tonnes) to bypass the dam in about four hours. A shiplift, or ship elevator, raises or lowers vessels of up to 3,500 tons (3,000 tonnes) in about 30 minutes.

WHAT IS A MEGAWATT?

⚡ A megawatt, MW (1 million watts), is the measure of the electrical energy in a generating system at any one time (in the way that gallons or liters are measures of the amount of fluid in a container).

⚡ A megawatt-hour, MWh, is a measure of this energy use over a one-hour period.

⚡ The Three Gorges Dam's output is 22,500 MW —that's how much electrical energy is in its generating system at any one time.

⚡ One MW can supply between 500 and 1,500 typical homes in a developed country.

MORE CLEAN ENERGY

Electricity is the world's favorite form of energy. It is transportable and portable. It can be converted into light, heat, movement, sound, and many other kinds of energy. Hydropower and wind power are two of the main ways that the weather can be used to generate renewable, relatively eco-friendly electricity. The weather provides other ways, too, such as wave power. Even methods for using the Sun's energy directly, as light or heat, depend on the weather for their success.

WAVE POWER

Weather-driven winds and water currents, assisted by tides due to the Moon's gravity, create waves. Their moving water is a source of energy that could be converted into electricity. Inventors and engineers have been trying to achieve this for more than a century, using many and varied devices and technologies. The great majority have failed. The main problem is the ferocious power of storm-generated waves that can easily smash and wreck the generating devices, coupled with the corrosive nature of seawater.

SEA SNAKES

The first large-scale wave project, Aguçadoura Wave Farm off the coast of northern Portugal (below), used floating snake-like generators. The wave undulations at the joints generated 2.25 MW of electricity. It first produced electricity in 2008 but closed a few months later due to technical problems. The machines were eventually scrapped.

SALTY TO FRESH

Hot, dry climates have plenty of sunshine but often lack fresh water. Desalination is the process of removing salt from seawater to make fresh water. Some desalination methods use a lot of electricity or even fossil fuels, but there are more sustainable methods. This small solar-powered desalination plant is in Madagascar. Electricity from solar panels powers "reverse osmosis." This means that seawater is pushed by an electric pump through a membrane that's covered in tiny holes, leaving the salt behind.

DIRECT SOLAR 1: HEAT

Solar thermal energy systems convert the Sun's heat, or IR (infrared) rays, into electricity. With current technology, they can only work in places with exceptionally warm, sunny climates—mainly hot deserts. In many designs, the Sun's IR is reflected by banks of moving mirrors, which track the Sun, onto a central tower. Here, the rays melt a substance, such as sodium, which stores the heat and uses it gradually to boil water and drive steam turbines, even without sunshine. The Noor III Ouarzazate Solar Power Station (below), in the Sahara Desert in northeast Morocco, has this kind of design.

> PROS

- ☀ Renewable energy.
- ☀ Once installed, little maintenance is needed.
- ☀ No noise or other disturbance.
- ☀ Once constructed, no greenhouse gases or other emissions.

> CONS

- ☀ Costly to build.
- ☀ Dependent on location and weather conditions.
- ☀ Limited energy storage in overcast conditions.
- ☀ Unsightly and spoil scenery.
- ☀ The super-hot beams to the tower can harm wildlife such as birds and insects.

DIRECT SOLAR 2: LIGHT

Solar panels are photovoltaic (PV), meaning they convert light rays to volts of electricity. Solar panels exploit the Sun's energy directly, but are dependent on the weather. Cloudy, overcast locations mean less sunlight and so less electricity. Most modern solar panels convert about 20–25 percent of light energy into electrical energy. Their use in solar farms is growing fast, and they now generate nearly one-thirtieth of global electricity. These solar panels are in Qinghai province, central China.

> PROS

- ☀ Renewable energy.
- ☀ Costs decrease as more produced.
- ☀ Little wear or maintenance.
- ☀ Once installed, no greenhouse gases or other emissions.

> CONS

- ☀ Depend on weather conditions.
- ☀ Only make electricity during sunlight hours.
- ☀ On farmland, they use up valuable crop-growing or grazing areas.
- ☀ Considered unsightly by some people, spoiling the scenery.

SALTY TO SALT

In sunny coastal areas, the Sun's heat and the winds it produces have been used for millennia to separate salt from seawater. The seawater is channeled into wide, shallow ponds and the water evaporates to leave behind sodium chloride—the substance we use for cooking and table salt. This is then harvested (shown here on a salt farm in Thailand) and purified. In centuries past, salt was important for preserving foods, especially meats, and was extremely valuable, though it is mainly used as a flavoring today and is much cheaper.

WEATHER-POWERED FUN

Moving air and water, caused by the weather, are a renewable, sustainable source of leisure, fun, excitement—and intense competition. Wind-driven activities range from windsurfing to racing multimillion-dollar yachts. Flowing water provides some of nature's greatest thrills, including white-water rafting and "shooting rapids" in kayaks.

AROUND THE WORLD BY WIND

One of the most grueling of all sports, physically and mentally, is long-distance ocean-yacht racing. Crews can be at sea for weeks on end. They must constantly check the weather, plotting a course that uses the wind but avoids storms and big waves, and keep the boat running well. Only the sturdiest, most advanced vessels, such as the one below, can take part in the Clipper Round the World Race, which takes 10 months and covers more than 45,000 miles (72,000 km).

BLOWING ALONG

Wind blowing across water propels a huge variety of watersports, from kitesurfing and windsurfing to many kinds of sailing. On land, wind can also power land yachts (below)—wheeled vehicles fitted with sails. Land yachting is best suited to shorelines where there are strong, steady winds and a large, flat expanse of beach left behind by the tide.

📷 A clipper yacht departs from Hainan at the start of Leg 8 of the Clipper Round the World Yacht Race in 2018.

UP AND AWAY

Wind-powered flight takes many forms, from paragliding and hang-gliding to gliding using full-size sailplanes. A paraglider (below) sits in a harness beneath a parachute-like device. Participants must use their skills to find warm, rising air currents—known as thermals and updrafts—to gain height and stay in the air for longer. Understanding the atmosphere and weather—for example, different kinds of clouds, how wind blows up slopes, and why land surfaces such as dark rocks heat the air above—is also useful.

FIGHTER KITES

Kite-flying can be a restful pastime—or a deadly serious contest. Kite-fighting has been popular in many Asian countries for centuries and is spreading to other regions. To play, each competitor flies a carefully designed kite on a single line.

Tightening or loosening the line allows the kite to rock and sway, as well as climb, dive, dash sideways, or even spin. The ultimate aim of the game is to cut through the line of an opponent, so the strings are coated with tiny pieces of sharp material, such as glass.

WINDY WORLD RECORDS

Windy weather can push sailing craft to amazing speeds.

- The fastest wind-powered water vessels are specialized racing catamarans with sails shaped like aircraft wings. In 2012, the *Vestas Sailrocket 2* (shown here) set a world sailing speed record of 75.2 mph (121.1 km/h).
- Top speeds in windsurfing and sailboarding can reach 60 mph (100 km/h).
- The most rapid land yachts can zip along at over 125 mph (200 km/h).
- The top hang-glider speeds over an official course are around 40 mph (65 km/h).

FORECASTING WEATHER

Predicting, or forecasting, the weather is a complex science, which has given rise to an immense industry worth more than $10 billion each year—and it is growing fast. So much of our daily lives, from travel, vacations, and sporting events, to choosing clothes each morning—and even being a space tourist—depend on the weather.

Meteorology is the scientific study of the atmosphere and weather in order to forecast it. More than half a million qualified meteorologists work around the world, but they are not the only people involved in weather prediction. The design, installation, and maintenance of weather stations and equipment employs engineers and many other specialists. Computer scientists develop programs to analyze the gigantic amounts of information gathered on the weather. TV presenters tell us what weather to expect. Also involved are climatologists, space experts, and many other scientists. With the challenge of climate change, and predictions of more extreme weather events, the need for fast and accurate weather forecasting will only increase.

"SPACE WEATHER"

Phenomena out in space can affect the weather on Earth. The Sun goes through an 11-year cycle during which the number of sunspots rises then falls. More sunspots lead to an increase in solar flares, which send large amounts of charged particles heading toward Earth. Earth's magnetic field protects us against these particles, although a large rise in sunspot activity can produce an increase in the Northern and Southern Lights (see page 45) and possibly some disruption to radio transmissions and power grids. Satellites are used to identify the details of this process.

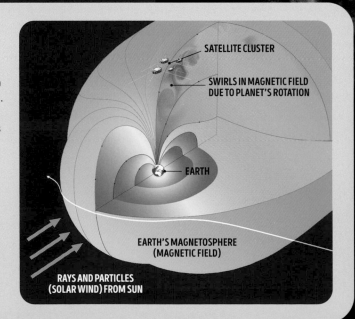

SATELLITE CLUSTER

SWIRLS IN MAGNETIC FIELD DUE TO PLANET'S ROTATION

EARTH

EARTH'S MAGNETOSPHERE (MAGNETIC FIELD)

RAYS AND PARTICLES (SOLAR WIND) FROM SUN

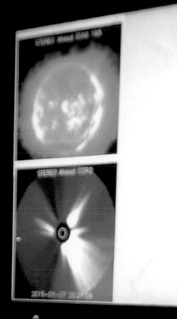
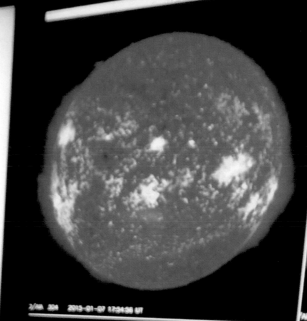
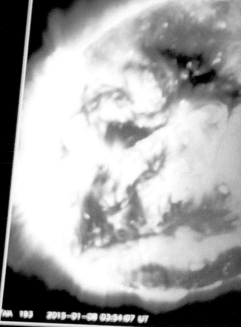
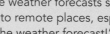

📷 A space weather forecaster monitors Sun activity at NOAA's Space Weather Prediction Center, Colorado.

FATAL MISTAKE

Accurate weather forecasts save lives. Guidance for people who venture into remote places, especially hills and mountains, includes, "Check the weather forecast!" Ignoring this advice can be deadly. Weather conditions can change in minutes. Those who are not prepared can suffer hypothermia (very low body temperature), frostbite, serious injury, and worse. And the people who help them, such as this air rescue team in Canada, face their own risks.

NATURE'S FORECASTERS

Long before the science of meteorology, people noted how the behavior of certain animals and plants could indicate the coming weather—foretelling a dry spell or heavy rain, for example, or the approach of a storm. As people began to keep farm animals and domesticate pets, such as cats and dogs, they noted more behavior related to the weather. Many of these stories about how animals predict weather became part of folklore. Some are supported by modern science, while others remain as unproven traditional stories.

FORECASTING FOLKLORE

There are many English-language sayings about forecasting and interpreting the weather.

▶ **When cows lie down, expect bad weather. They stand up in heat, to cool their udders.**
There is some rationale for this. Cows lose less body warmth when lying down, which would help in cold, windy weather. Similarly, standing up in hot weather helps air circulation around the legs, which would lessen the risk of overheating.

▶ **Plenty of fruits and berries are a sign of a hard winter to come.**
This saying is difficult to support. Lots of fall berries, for example, on brambles, holly, and rowan trees, are more a sign that the year had a fine spring and a warm but not too-dry summer.

▶ **Birds flying high, few clouds in the sky. Birds flying low, prepare for a blow.**
Winds are stronger at greater altitudes. In sunny, calm weather, birds can fly more easily at height. Also, small insects fly higher when it's hot, so birds such as swallows and swifts follow to feed on them.

THE CRICKET THERMOMETER

Chirping male crickets may not forecast weather, but some kinds can indicate the temperature, as discovered in 1897 by US scientist-inventor Amos Dolbear. He knew that the insects' bodies are usually at the temperature of their surroundings. As the temperature rises, their muscles work faster, so the crickets' chirps become more rapid.

Dolbear discovered that to figure out the temperature in °F, you must count the number of chirps in 14 seconds, then add 40. For °C, count the chirps in 25 seconds, divide by 3, then add 4.

EXAMPLE:
28 chirps in 14 seconds + 40 = 68°F (20°C).
50 chirps in 25 seconds ÷ 3 = 16.6 + 4 = 20.6°C (69°F).

A herd of migrating zebra in the Okavango region, southwest Africa

SHADOW OR NO SHADOW?

A North American tradition that probably came from Germany is Groundhog Day, on February 2 each year. The story goes that if a groundhog (below) comes out of its burrow on this day, sees its shadow, and goes back into its burrow, then it means that winter will continue for another six weeks. According to the theory, if there's a shadow, it must be sunny, with the weather characterized by high pressure, clear skies, low temperatures, and frosts, making it difficult for the groundhog to find food. If there isn't a shadow, then that means the conditions are cloudy and cool, indicating low pressure and an early, mild spring, so the groundhog begins to forage. Scientific checks do not support this belief, but people have lots of fun celebrating the custom.

FEATHERED WEATHER REPORTS

In 2013, in Tennessee, a flock of golden-winged warblers received tiny geolocator devices to record their migrations. Unexpectedly, one day the birds suddenly dispersed. The very next day, a massive supercell thunderstorm hit the area. Many similar stories tell how birds predict the weather—especially changes in air pressure— which affects their travels and feeding. Inside the ear, most birds have a special feature known as a paratympanic (or Vitali) organ. This is very sensitive to atmospheric pressure and is thought to act as their natural barometer-altimeter (see page 138).

ON THE MOVE

In 2015, in the Okavango region of southwest Africa, scientists fitted zebras with radio-tracking collars to follow their regular migration, which takes in more than 155 miles (250 km) from a dry area to fresh grassland. The zebras' movements were compared with satellite weather reports. The results showed how the herds timed their journeys, and quickly changed or even reversed their routes, according to weather conditions. Especially important was distant rainfall that encouraged new plant growth, with herds reacting days, or sometimes even hours, in advance of the coming rains.

EARLY WEATHER REPORTS

Many ancient civilizations forecasted the weather, from Mesopotamia, Egypt, India, and China, to Greece, Rome, and Mesoamerica. However, many of these forecasts were part of a wider system of beliefs, woven into tradition and religion. For example, high priests would declare that floods and storms were sent by sky gods to punish those who did not worship them or offer valuable gifts—often gifts for the high priests themselves. Meanwhile, ordinary people looked at the clouds, saw the wind direction, noted if they felt warmer or cooler, and made informed guesses about tomorrow's weather.

📷 Egyptian Pharaoh Akhenaten and his family worshipping the god Aten, depicted as a disc of the Sun, in the 14th century BCE

ANCIENT TIMES

▶ In ancient Egypt, 3,000 years ago, people believed the weather predictions of their pharaoh. After all, he was supposed to be in direct contact with the Sun god.

▶ More than 2,000 years ago in ancient China, clouds were central to forecasting, with sayings such as "Clouds the shape of castles foretell heavy rain." Such towering clouds are known today as cumulonimbus.

▶ In the 1400s, the Inca people of South America would study the stars before sowing the seeds for next year's crop. A clear, sharp view meant forthcoming good weather and a plentiful harvest. Blurred, hazy stars indicated bad weather, and sowing dates were changed.

EARLY WEATHER BOOK

One of the earliest accounts of weather was written more than 2,300 years ago by Greek philosopher and scientist Aristotle. His *Meteorologica* contained what we now know to be excellent insights, but also strange errors. Using the traditions of the Greek gods, Aristotle described storms as battles between "good" and "evil" winds. He also said that the Sun, by its "agency," made the sweetest (that is, non-salty) water rise up into the air, dissolve as vapor, then be condensed by cold and return to the Earth—what we now call the water cycle (see pages 26–27).

Italian mathematician and physicist Evangelista Torricelli with his barometer

WEATHER REPORTS

Weather forecasting for the general public began in the 1860s, mainly in the United Kingdom. Before then, gathering weather information from far and wide took so long, even by horseback, that any predictions were already out of date by the time people had a chance to read them. In 1860, the Royal Navy's vice-admiral Robert Fitzroy, who was worried about ship losses in bad weather, organized information to be collected by the newly invented electric telegraph system. He produced charts that he called "forecasts." Daily forecasts began the next year in the *Times* newspaper (left), and many other countries soon followed.

BEGINNINGS OF METEOROLOGY SCIENCE

From the 1400s onward, a revival of learning and art, which came to be known as the Renaissance, began in Europe. It saw the emergence of many new scientific ideas. Inventors came up with new devices that could accurately measure natural phenomena. The first thermometers to record temperatures were used in the 1600s. They were followed in the early 1700s by a more reliable version, devised by the German scientist Daniel Fahrenheit (after whom the temperature scale is named). His thermometer used mercury in a sealed glass tube and remained common into modern times. Measurement of air pressure began with the invention of the barometer by Italian physicist Evangelista Torricelli in 1643.

RADIO TAKES OVER

Today, weather forecasts race around the world at lightning speed by means of radio waves, fiberoptics, and similar technology. Radio waves were first used to send out regular forecasts in 1902, mainly to ships at sea. The technology had been created by the Italian inventor Guglielmo Marconi. Three years later, ships themselves began sending radio signals about their weather conditions at sea to weather centers on land.

Malin Head in County Donegal, Ireland, where the Marconi company sent the first commercial message by radio to a ship at sea

The Norwegian Cyclone Model combined data about surface and high-altitude winds to explain how a cyclone, or low-pressure area, forms and begins to rotate.

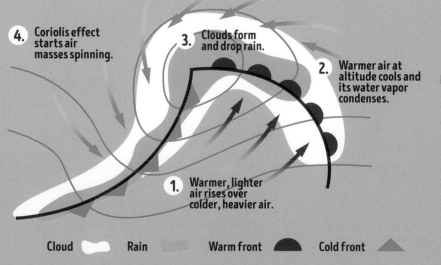

4. Coriolis effect starts air masses spinning.

3. Clouds form and drop rain.

2. Warmer air at altitude cools and its water vapor condenses.

1. Warmer, lighter air rises over colder, heavier air.

Cloud   Rain   Warm front   Cold front

THE IMPORTANCE OF ALTITUDE

Before the 20th century, most meteorologists focused on low-level weather. As science progressed, they began to think about how upper layers of the atmosphere were also involved. This was helped by measurements gathered by kites, balloons, and planes. From about 1910 in Norway, Henrik Mohn, Jacob Bjerknes, and their colleagues suggested how areas of swirling low pressure, called cyclones, begin when a mass of lighter, warmer air rises over colder, heavier air, then slides along it and begins to spin, due to the Coriolis effect (see page 109). This early work became known as the Norwegian Cyclone Model. It led to much improved forecasting and modern computer modeling.

WEATHER STATIONS

Places where weather measurements, such as wind speed, amount of rainfall, and hours of sunshine, are taken are known as weather stations. They can be as simple as a home-made rain gauge in a backyard, or as complicated as a whole building crammed with recording devices, electronics, computers, and meteorologists. Usually, official weather stations are set up and monitored by a country's national meteorological organization. There are also millions of unofficial amateur and hobby weather stations in almost every possible place, from homes, schools, and farms to city skyscrapers.

WEATHER EVERYWHERE

Officially recognized weather stations are spread across the land and sea, in every kind of habitat and landscape —mountaintops, hillsides, valleys, deep canyons, deserts, swamps, lakes, coasts, and the open ocean. They are carefully sited to provide the greatest range of information for different weather and climate conditions. The weather station on the left is on a mountaintop in Italy.

ANYONE CAN HAVE ONE

Modern electronics have made weather stations far less costly than before. Many schools and colleges have them for use in their courses, like the one shown below. Students can get involved in everything from making simple weather observations to learning about the physics of the atmosphere, as well as specialized meteorology and climatology. These stations can be linked via the Internet to local and national weather centers, providing valuable data that is fed into the forecasting system.

MR. STEVENSON'S SCREEN

In the 1860s, Thomas Stevenson, a Scottish engineer and meteorologist, designed a box-like structure with a roof, slatted sides, and no floor, which was mounted on a post or legs. His idea was to provide a standard shelter for weather devices, such as thermometers and barometers, which were fixed to the inside walls. The simple design could be made and used anywhere, so that measurements could be compared accurately. It was painted white to reflect direct sunlight and shield the interior, so the thermometer inside measured the true air temperature. The Stevenson screen (above) became an accepted design for small-scale weather stations around the world.

FULLY AUTOMATED: THE AWS

Traditional weather stations record measurements on mechanical devices. Someone has to visit regularly, take readings, reset the devices, and check all is working well. Most modern versions are automatic weather stations (AWS), which are electronic and self-contained with very little maintenance required. Their batteries are charged by solar panels or small wind turbines. They store information in a data logger and send it via radio connections, satellite links, or a cell phone network straight to weather centers. This AWS is at Spitsbergen, Norway, in the Arctic.

WEATHER AT SEA

Ships and boats detect the weather for their own safety and to send data to weather centers. They also receive forecasts so they can prepare for any problems. Usually, the meteorological devices on board are part of the ship's general navigation systems. The Swedish ice-breaking ship *Oden* (below) is a research vessel that has traveled the globe recording wind, rain, snow, waves, currents, ice, marine life, and makeup of the ocean and atmosphere, as well as hundreds of other measurements.

NATIONAL WEATHER

Nearly all countries have an official weather organization. For example, the National Weather Service (NWS) is the agency of the US federal government that provides weather forecasts, extreme weather warnings, and other weather-related information. Other similar national organizations include:

- ☁ **UK** – Met Office
- ☁ **FRANCE** – Météo-France
- ☁ **AUSTRALIA** – Bureau of Meteorology
- ☁ **GUATEMALA** – National Institute for Seismology, Vulcanology, Meteorology and Hydrology, INSIVUMEH (Instituto Nacional de Sismología, Vulcanología, Meteorología e Hidrología)
- ☁ **INDONESIA** – Meteorology, Climatology, and Geophysical Agency, BMKG (Badan Meteorologi, Klimatologi, dan Geofisik)

These, and more than 180 others, are members of the World Meteorological Organization, WMO (see page 82).

WEATHER MEASURERS

More than 50 kinds of meteorological instruments measure and record almost every aspect of the weather. Some are familiar, like thermometers for measuring temperature, and the spinning cups of the anemometer for calculating wind speed. Others are less well known, such as the pyranometer that detects the power, wavelength, and direction of the Sun's different rays. Many of these devices have both a traditional mechanical design and a modern electronic version (see page 138).

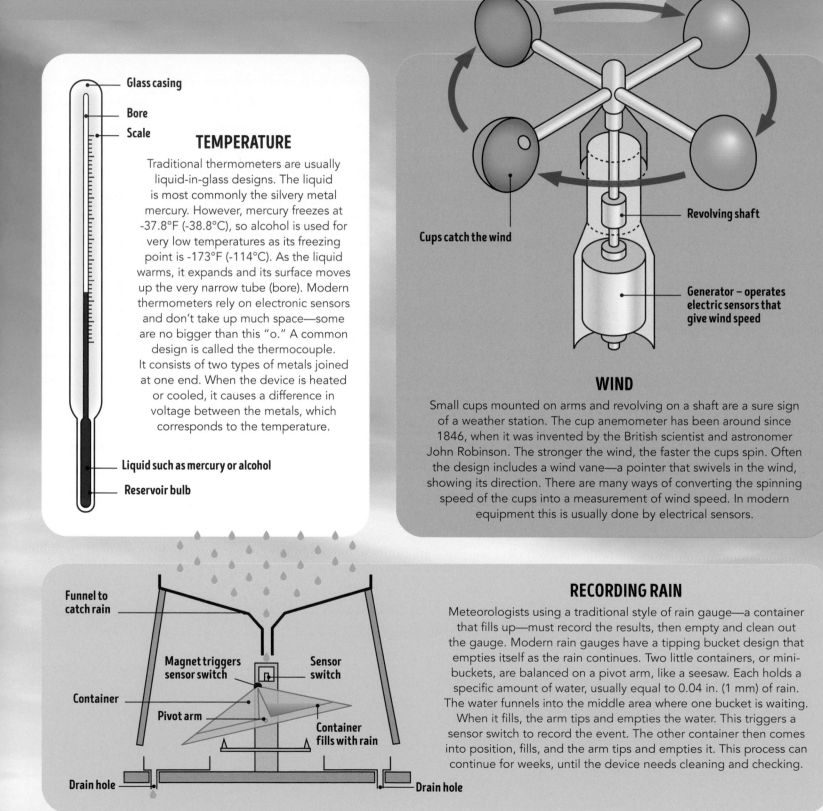

Glass casing

Bore

Scale

TEMPERATURE

Traditional thermometers are usually liquid-in-glass designs. The liquid is most commonly the silvery metal mercury. However, mercury freezes at -37.8°F (-38.8°C), so alcohol is used for very low temperatures as its freezing point is -173°F (-114°C). As the liquid warms, it expands and its surface moves up the very narrow tube (bore). Modern thermometers rely on electronic sensors and don't take up much space—some are no bigger than this "o." A common design is called the thermocouple. It consists of two types of metals joined at one end. When the device is heated or cooled, it causes a difference in voltage between the metals, which corresponds to the temperature.

Liquid such as mercury or alcohol

Reservoir bulb

Cups catch the wind

Revolving shaft

Generator – operates electric sensors that give wind speed

WIND

Small cups mounted on arms and revolving on a shaft are a sure sign of a weather station. The cup anemometer has been around since 1846, when it was invented by the British scientist and astronomer John Robinson. The stronger the wind, the faster the cups spin. Often the design includes a wind vane—a pointer that swivels in the wind, showing its direction. There are many ways of converting the spinning speed of the cups into a measurement of wind speed. In modern equipment this is usually done by electrical sensors.

Funnel to catch rain

Magnet triggers sensor switch

Sensor switch

Container

Pivot arm

Container fills with rain

Drain hole

Drain hole

RECORDING RAIN

Meteorologists using a traditional style of rain gauge—a container that fills up—must record the results, then empty and clean out the gauge. Modern rain gauges have a tipping bucket design that empties itself as the rain continues. Two little containers, or mini-buckets, are balanced on a pivot arm, like a seesaw. Each holds a specific amount of water, usually equal to 0.04 in. (1 mm) of rain. The water funnels into the middle area where one bucket is waiting. When it fills, the arm tips and empties the water. This triggers a sensor switch to record the event. The other container then comes into position, fills, and the arm tips and empties it. This process can continue for weeks, until the device needs cleaning and checking.

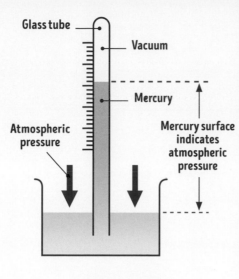

Glass tube
Vacuum
Mercury
Atmospheric pressure
Mercury surface indicates atmospheric pressure

AIR PRESSURE: 1

Barometers come in many shapes and sizes. The Torricellian design (see pages 17, 133), named after its inventor, consists of an upside-down glass tube with its open lower end in a container of mercury. The space above the mercury is a vacuum (totally empty, even of air). The surrounding atmospheric pressure pushes on the surface of the mercury in the container and forces it up in the tube. If the barometer is at the standard atmospheric pressure experienced at sea level, then the mercury will rise to a height of about 30 in. (760 mm). As the surrounding atmospheric pressure rises or falls, so does the mercury in the tube.

A traditional wooden barometer

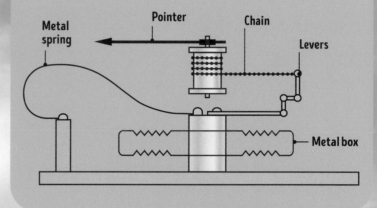

AIR PRESSURE: 2

An aneroid (or dial) barometer uses a sealed, slightly flexible metal box to measure air pressure (above). As atmospheric pressure changes, it either squashes the box or allows it to expand. Springs and levers attached to the box measure the movement and display it using a pointer that moves around a dial. Up-to-date digital barometers, smaller than a garden pea, measure how atmospheric pressure deforms a tiny sheet of material and changes an electric current.

Metal spring
Pointer
Chain
Levers
Metal box

HUMIDITY

Hygrometers measure humidity—the amount of water vapor in the air. An old design used long hairs (human or animal) that were slightly stretched and attached to a pointer. In damp conditions the hairs would absorb water vapor and become a tiny bit longer. As they dried out, they shortened, moving the pointer. Some modern digital hygrometers have a surface covered with a special chemical that changes its resistance to an electric current according to how much moisture it absorbs from the air. Another design measures how electricity passes through the air itself—moist air conducts current better.

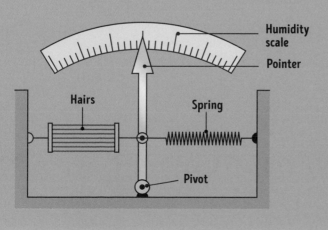

Humidity scale
Pointer
Hairs
Spring
Pivot

COMPACT DIGITAL WEATHER STATION

A modern compact digital weather station can send weather forecasts, as well as readings for temperature, air pressure, humidity and other conditions, direct to your phone. It can also display the time, day, and date, and has links to other sensors and the Internet. Some also play music!

MORE WAYS TO MEASURE WEATHER

Standard meteorological devices, like those on the previous pages, are familiar in homes, schools, offices, factories, farms, and many other places. There are also more specialized, technical, and costly instruments, usually found only in larger weather stations or places where weather has an enormous influence, like airports and seaports. Some have traditional designs based on mechanics, physics, and chemistry, like the Campbell-Stokes sunshine recorder (see page 85), while others use only the latest electronics and computer technology.

CLOUDS

The ceilometer (below) measures the "cloud ceiling," which is the distance from the ground to the cloud base, and also the cloud's thickness from base to top. Some give a general visibility reading, too. It is a common instrument at airports. Most designs shine narrow laser beams (or other types of light rays) upward and use photodetectors (small, light-sensitive electronic circuits) to identify reflections and how the rays are scattered. The time from sending a beam to detecting the reflections can be used to measure the cloud's height.

SNOW

A pole-mounted ultrasonic snow depth sensor, such as the one in this picture, works by sending very high-pitched (ultrasonic) sound waves down to the ground and timing the returning echoes. Knowing the speed of sound gives the distance from the sensor to the ground. When snow falls, the ultrasonic waves bounce off its surface and return more quickly. A wireless link can then transmit this information to where it's needed, such as centers in charge of ski areas, roads, or railroads. Some sensors are accurate to less than 0.04 in. (1 mm).

ATMOSPHERIC PARTICLES

Dust, mist, fog, smog, soot, smoke, vehicle exhausts—these are all types of particles floating in the air. They indicate the weather conditions and also affect them. Devices called airborne particle detectors, aerosol photometers, and nephelometers collect and analyze the particles. Most use laser light that shines through a small chamber containing the air and its particles. Photodetectors measure how the light is reflected, scattered, or blocked, to determine each particle's size and shape. Some of these devices are hand-held, like the one on the left, while more complex and accurate versions are often installed by roadsides to monitor traffic fumes.

SOLAR ENERGY

A pyranometer (below) measures solar irradiance—the energy of waves and rays coming from the Sun. Using various electronic sensors, most designs can distinguish between direct sunlight, sunlight passing through clouds, and sunlight scattered off clouds, as well as similar features of the Sun's infrared and ultraviolet waves. A pyranometer can also detect the angles at which these forms of energy come from in the sky. Pyranometers are particularly useful at solar power plants, where the amount of the Sun's energy reaching the area can be compared with the amount of electricity being produced.

YET MORE WEATHER GADGETS

▶ Visiometer – measures visibility, or clarity, through the air (or water).

▶ Lightning detector – warns of approaching storms and electrical discharges.

▶ Disdrometer – measures drop size, distribution, and speed of falling precipitation, from light drizzle to heavy rain and hail.

▶ Wind profiler – radio or sound waves measure wind speed at various heights.

▶ Pan evaporator – measures the amount of water needed to replenish a wide pan to keep its depth constant, showing the amount lost as it evaporates.

▶ LIDAR (Light Detection and Ranging) – uses pulses of laser light to bounce off objects and detect the reflections in order to measure distance. Sometimes called 3D laser scanning, it has many uses in meteorology, such as studying clouds and rain, mapping the Earth's surface, and measuring wave height, length, and speed at sea.

WEATHER RADAR

Most weather radar systems consist of a network of large domed radar equipment mounted on towers. This sends out microwave signals and picks up their reflections (echoes), to detect and track rain, snow, and other precipitation, as well as wind. The US system, NEXRAD, has 160 radar towers spread across the country. However, there are gaps in the coverage in some particularly mountainous or rural areas. If severe weather begins in these places, it might not be detected for hours, even days, until it moves in to the next coverage area.

◎ A weather radar tower at sunset in Saskatchewan, Canada

WEATHER SATELLITES

Every second of every day, satellites orbiting Earth collect massive amounts of weather and climate information. Their main job is to watch the weather and measure the climate, although some have other capabilities, too. Weather satellites are usually launched and managed by a particular country, but the information they collect is shared around the world. There are many other satellites in space, which have different jobs, such as searching for new stars and planets, carrying radio, telephone, and TV signals, or "spying" on other countries—but they can also collect weather and climate information.

THE METEOSAT SYSTEM

Different nations and groups operate their own weather satellites within a global system. For example, EUMETSAT (European Organisation for the Exploitation of Meteorological Satellites) controls the Meteosat series of satellites, the first launched, in 1977. Several have been developed over the years, which have become increasingly technologically advanced. The latest, Meteosat-11, should stay in service until well past 2030. These types of satellites send the information they gather as codes of radio signals to receiving stations on Earth. They receive their instructions from Earth in a similar way.

WHAT WEATHER SATELLITES DETECT: 1

Most weather satellites have camera-like electronic sensors, or imagers. These can pick up two different kinds of energy, known as electromagnetic radiation. The first kind is visible light, which we can also see. The resulting images, such as the one on the right of northeastern Spain, show familiar features such as clouds, clear skies, seas, lakes, mountains, forests, cities, smoke from large fires, and types of pollution, such as smog and dust, as well as floods and snow.

NORTH AMERICA

Atlantic Ocean

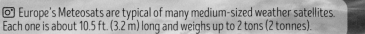
📷 Europe's Meteosats are typical of many medium-sized weather satellites. Each one is about 10.5 ft. (3.2 m) long and weighs up to 2 tons (2 tonnes).

WEATHER SATELLITE ORBITS: POLAR

There are two main kinds of weather satellite orbits—polar and geostationary. A polar orbit goes over, or nearly over, each pole, at heights of between 125 and 625 miles (200 and 1,000 km) and takes less than two hours. The satellite always keeps to the same path as the Earth rotates beneath. So the satellite covers the planet's whole surface in 24 hours and goes over the same spot on the surface at the same time each day.

WEATHER SATELLITE ORBITS: GEOSTATIONARY

A geostationary orbit takes place directly above the Equator at a distance very close to 22,236 miles (35,786 km). The time taken for this orbit is 24 hours, which is the same amount of time it takes Earth to rotate once. So, when seen from Earth, the satellite seems to always hover at the same place in the sky. There are about 400 geostationary satellites in orbit. By international agreement, they are spaced well apart to lessen the risk of collision.

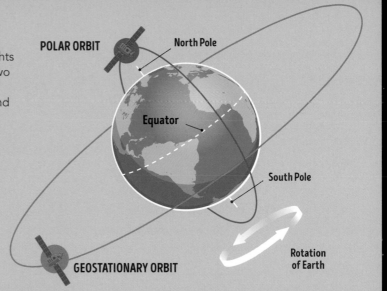

POLAR ORBIT

North Pole

Equator

South Pole

GEOSTATIONARY ORBIT

Rotation of Earth

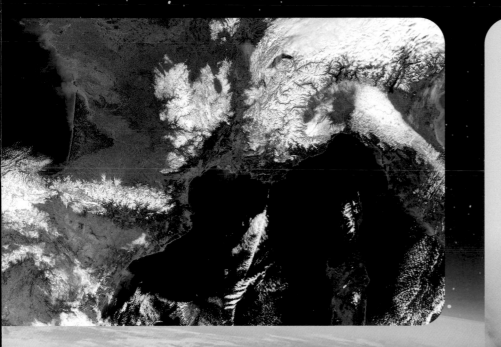

NOT QUITE IN SPACE

About 1,000 weather centers around the world release weather balloons up into the sky— sometimes up to four a day! A typical balloon is very stretchy and contains hydrogen gas, which is lighter than air. As the balloon rises, the air pressure drops and the balloon expands (gets bigger). Inside is a miniature weather station and transmitter called a radiosonde, which measures altitude, air pressure, humidity, temperature, wind speed and direction, and geographical position, as well as more specialized information, such as ozone levels. Eventually, after around two hours, at a height of about 12–25 miles (20–40 km), the balloon bursts and the radiosonde parachutes back to Earth.

Atlantic Ocean

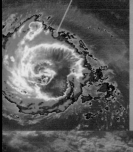

HURRICANE FLORENCE

WHAT WEATHER SATELLITES DETECT: 2

The second type of electromagnetic radiation detected by weather satellites is infrared, or IR, which is invisible to the naked eye. It's detected by devices known as radiometers. They can provide detailed information about cloud temperature patterns that signify storms— as this image of Hurricane Florence over the Atlantic Ocean in 2018 shows. The colored areas indicate the temperature at the top of the clouds. Although we usually associate red with heat, in these images, red is coldest, followed by yellow, green, blue, and black. These colors enable forecasters to figure out if the storm is developing or slowing down. When cloud top temperatures get colder, it means that they're getting higher into the atmosphere. This means the "uplift" of warm, moist air is stronger, and the storm is powering up.

A weather balloon is launched in Colorado in 2019.

CASE STUDY:

GOES WEATHER SATELLITES

The USA launched its first Geostationary Operational Environmental Satellite, GOES-1, in 1975. Since then, another 17 have gone up into space. At present, four are active. These satellites send vast amounts of weather-related information to Earth. This is shared around the world, helping countries make all kinds of forecasts, from long-term climate change predictions to short-term warnings for hurricanes, floods, snowstorms, and droughts. The whole GOES system of satellites, ground stations, weather centers, and forecasting is part of an organization known as NOAA—National Oceanic and Atmospheric Administration.

LOCATION

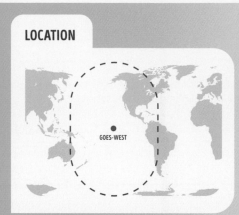

GOES-WEST

GOES-17 IS GO

Launched in 2018, GOES-17 (left) is the second of a new series of GOES satellites with improved weather-watching abilities. When it arrived in its westerly position, it was also given the name GOES-West. Sitting above the Equator approximately in the middle of the Pacific Ocean, it can watch almost half of the globe—much of North America and northwest South America as well as New Zealand and eastern Australia. A ship in the Pacific, on the Equator, 3,750 miles (6,000 km) west of Ecuador, would see GOES-17 directly overhead, not moving.

TRACKING CLOUDS

GOES satellites record images at various time intervals, from a few seconds to days apart. This image shows a collection of clouds passing eastward over the US Hawaiian Islands in the Pacific Ocean— the islands are shown in outline. It was part of a series of photographs, taken minute by minute on November 15, 2018, at the tail end of a particularly fierce hurricane season, to monitor how the clouds developed, moved, and faded.

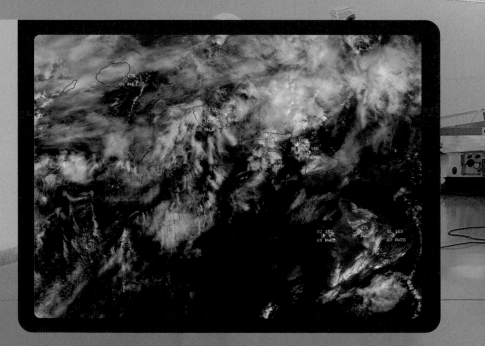

ONBOARD GOES-17 GOES-WEST

The main image sensor on the new GOES satellites is the Advanced Baseline Imager (ABI). This picks up 16 sets of visible and infrared rays. It monitors many aspects of the weather and environment: clouds; winds; land and water surface temperatures; ocean currents; river-flow rates; fires; clouds of smoke ash and gases from volcanoes; air quality; and even the health of plants and farm crops. Another important instrument onboard is the Geostationary Lightning Mapper (GLM), which detects all forms of lightning and shows where and how storms are developing.

BUILDING AND TESTING

The latest generation of GOES satellites are large—20 ft. (6 m) long, weighing almost 3 tons (3 tonnes). Here, GOES-17 is being checked before it is launched at Cape Canaveral, Florida. Despite such lengthy and detailed testing, problems sometimes occur. Soon after the launch of GOES-17, its main device, the ABI (see above), developed a fault that could not be repaired. The satellite's expected lifespan was reduced from 15 years to about five. This is an expensive problem. The latest series of four GOES-sats have a combined cost of $12 billion.

High levels of cleanliness must be maintained at all times. Workers are all suited and masked to prevent contaminating the delicate instruments of the satellite before it is launched.

WEATHER SUPERCOMPUTING

Every second, gigantic amounts of information pour into weather centers from thousands of weather stations, radars, weather balloons, ships, aircraft, and satellites all around the globe. Most of this information is in the form of digital data, which is analyzed by meteorological supercomputers. Their first task is to process it all into various categories such as rainfall, wind speed, and temperature, linked to location. This information is then fed into computer programs called weather prediction models, which compare and analyze the information to provide a weather forecast.

ROOM FULL OF DIGITAL POWER

Even though many of the modern digital devices and computers we use in our day-to-day lives can be incredibly small, weather supercomputers need to be so powerful that they fill large rooms. In 2019, the French weather service Météo-France bought two vast new computing machines (shown here) at a cost of more than $47 million. They increased the computing power of the service by five times. Each of the new supercomputers can handle more than 10 million billion calculations every second.

METEO MODELS

To analyze incoming weather data and turn it into forecasts, computers use very complex programs called weather prediction (or meteorological) models. These not only predict the likely weather over a range of time scales, from minutes to months, they also constantly compare past weather predictions at a certain location with what actually happened there. The accuracy of these forecasts is then fed back into the system to improve predictions. All this happens on a continual basis.

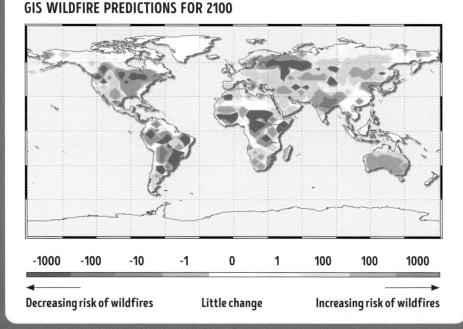

GIS WILDFIRE PREDICTIONS FOR 2100

-1000	-100	-10	-1	0	1	100	100	1000

← Decreasing risk of wildfires   Little change   Increasing risk of wildfires →

MODELING EXTREMES

As well as predicting the general weather and climate, there are computer models to forecast particular aspects in greater detail, such as rainfall, cloud cover, and risk of wildfires. Many of these use GIS, which stands for Geographic Information System. GIS analyzes all kinds of data, not only from weather and climate but also from industry, cities, and changes in farming. It picks out the particular information needed, for example, to forecast wildfires. It then runs the data through a specialized computer model and presents it as maps, charts, tables, and lists for regions around the world. As shown in the above image, GIS wildfire models predict that, with climate change, wildfires may become more common in some areas yet less frequent in others.

COMPARING COMPUTER MODELS

There are more than 200 major computer models used to predict weather and climate. These usually give similar, but not identical, forecasts. The map directly below shows some predicted paths for 2017's Hurricane Irma over the southeast US, with the calculated route of each model in a different color. Some models suggested Irma would swing northeast, out over the Atlantic. Others calculated Irma would head northwest, inland—which turned out to be its actual route, shown on the map underneath.

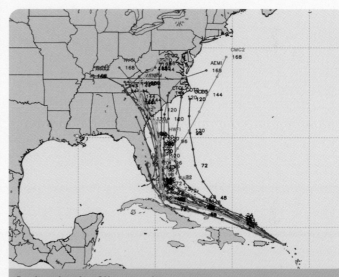

Predicted paths of Hurricane Irma

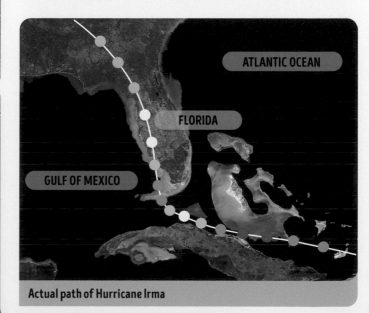

ATLANTIC OCEAN

FLORIDA

GULF OF MEXICO

Actual path of Hurricane Irma

BIG BUSINESS

Collecting and modeling weather and climate data, and making forecasts, are huge worldwide activities—and also big business. Many countries have national organizations (see page 135) that cooperate to share their weather information. Some of this is supplied to news companies—perhaps for free or maybe for payment—who can then provide forecasts for everyone along with the rest of the news. Specialized forecasters can also take the main weather data and analyze it in more detail to create predictions for a particular user—for example, accurate wind speeds at different heights for an airline. Each extra level of detail in a forecast usually costs more money. Specialized forecasting is worth more than $10 billion a year and growing fast.

A meteorologist at the National Weather Center Forecast Office in Oklahoma

"HERE IS THE WEATHER FORECAST"

Weather forecasts are available almost anytime, anywhere—on cell phones, tablets, computers, television, radio, the Internet, and in newspapers. These predictions are the result of a huge amount of work, starting with the collection of "raw" weather data, which is then processed through computer models and finally presented in a form that can be understood by a range of different people—from those looking for a quick daily forecast to people who need more detail and background, such as farmers or anglers.

RED SKY AT NIGHT

Long before the science of weather forecasting had been developed, humans relied on their experience to predict the weather—and there is often scientific evidence to back up these traditional forecasts. The old saying, "Red sky at night, shepherds' delight," was used to help anyone who spent much of their time outside plan for the next day. A red sky at night indicated that a day of fine weather was expected. Scientifically, a red sky appears when dust and small particles are trapped by high pressure in the atmosphere. This scatters blue light, leaving only red light to give the sky its glowing appearance. As high pressure is associated with fine, settled weather (see page 17), it means that the next day will usually be dry and pleasant.

TRICKY TO UNDERSTAND

When weather maps were first published in newspapers, you needed an expert eye to understand them. This example from an old 19th-century newspaper shows a mass of lines and shapes, which wouldn't have made much sense to most people, over a map of the USA. It is similar to charts we see today and gives lots of weather information, such as wind direction and strength, and whether there is high or low pressure. The key in the bottom corner shows the weather conditions for the towns and cities on the map, while the dashed lines indicate temperature in degrees Fahrenheit.

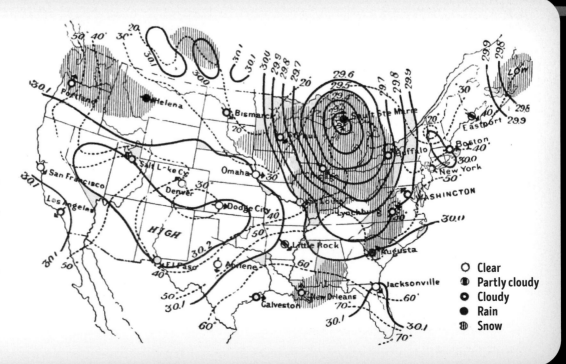

○ Clear
◑ Partly cloudy
◎ Cloudy
● Rain
❊ Snow

GRAPHICS AND ICONS

As with so much information, the main way of presenting a weather forecast is visual. Meteorologists and graphic designers have devised sets of symbols, or icons, which can be understood by anyone, whatever spoken or written language they use.

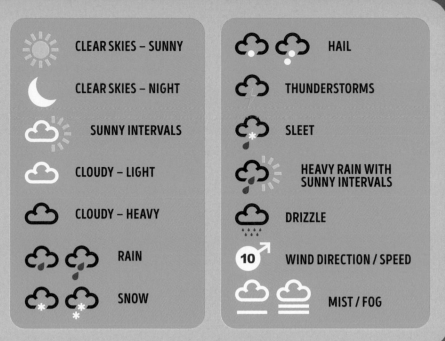

CLEAR SKIES – SUNNY

CLEAR SKIES – NIGHT

SUNNY INTERVALS

CLOUDY – LIGHT

CLOUDY – HEAVY

RAIN

SNOW

HAIL

THUNDERSTORMS

SLEET

HEAVY RAIN WITH SUNNY INTERVALS

DRIZZLE

10 WIND DIRECTION / SPEED

MIST / FOG

THE WEATHER MAP

Synoptic—or general—weather maps like this one show the current weather in an area. Lines and shapes on the map indicate various weather features, which are meant to convey to the reader what is happening. See below for explanations of what the symbols mean.

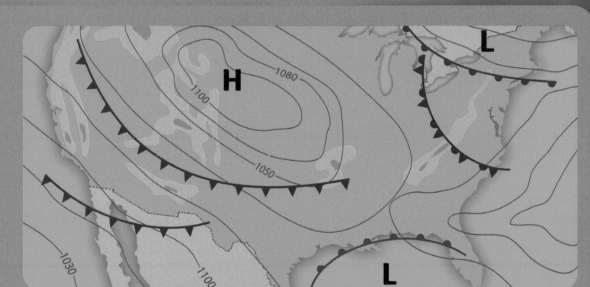

COLD FRONT
Triangles

This is where a mass of cold, dense air meets a mass of warmer, lighter air and pushes beneath it. The rising warm air may cause thunderstorms. As the cold front arrives, it brings gusty winds, and temperatures fall.

OCCLUDED FRONT
Both half-circles and triangles

This forms when a warm front gets caught between two cold fronts. Cold fronts travel faster than warm fronts, so they come together and force the warm air higher, causing clouds and rain, followed by drier weather.

WARM FRONT
Half-circles

Here, a mass of warm air pushes into a mass of cooler air and rises over it. This brings a wide area of rain, then low cloud and drizzle to the area below. As it passes, the temperatures rise.

ISOBARS

An isobar is a line that joins places with the same air pressure. When these lines are close together, it means that air pressure changes over a short distance. Air moves from high to low pressure, so the closer the lines, the quicker the air moves, meaning stronger winds.

RELYING ON THE FORECAST

Weather forecasts affect our lives in many different ways, from helping us choose what clothes to wear to deciding whether an outdoor event can take place safely. Extreme weather predictions avert disasters, prevent damage costing millions of dollars, and save lives. Many industries and occupations rely greatly on weather forecasts, from farming, aviation, and construction to outdoor sports and leisure activities—but even with all the money and technology behind them, weather forecasts are still sometimes wrong, because there are just so many different variables to consider.

CALL OFF THE GAME?

Outdoor sports are affected in different ways by weather forecasts. Soccer games can take place in most conditions, but snow on the field can be very tricky. The white lines on the field must be visible—and the ball, too! When snow is expected, the usual white ball is replaced with a brightly colored one, as seen here in this snowy 2016 game in Germany. Some stadiums have under-soil heating, which can be activated when snow and ice are forecast—but the stadium itself could still be declared unsafe due to the risk of people slipping. An accurate forecast is essential so the game can be called off in good time, if necessary, saving fans the trouble and cost of coming to the venue.

SAFE JOURNEYS

Weather forecasts are an essential resource for aviation. Knowing the predicted speed and direction of the wind allows aircraft pilots to fly around bad weather, or to save on fuel by avoiding a strong headwind. Landing in a crosswind—one that blows from the side—presents many challenges. But by using weather information provided by air traffic controllers, pilots can keep the aircraft pointed into the wind—where it's safest—until it's ready to touch down on the runway (as shown below). If severe crosswinds are forecast, the plane may be sent to land at another airport.

BLOWING IN THE WIND

Outdoor music festivals and similar events attract hundreds of thousands of people—but it pays to check the weather forecast before attending. Visitors to the UK's Glastonbury Festival in 2005 (below) were swamped by mud and rain, and the high winds blew many of the tents into the air.

GROWING SEASONS

Agriculture is organized around weather forecasts. Farmers need to have accurate predictions for rain, sunshine, dry spells, and wind, so they can plan activities like plowing, planting, pest-spraying, and harvesting (above)—and prepare for bad weather. If severe weather is due, livestock may need to be brought in to shelter ahead of time. Accurate forecasts are particularly vital around harvest time. Unexpected storms could result in the loss of the entire crop, costing immense amounts of money and causing possible food shortages—even famine.

PREPARE FOR PROBLEMS

Large construction projects are especially at risk from extreme weather. When storms are expected, workers need to check the building site to make sure that everything is fastened down safely. If the checks are not completed properly or the storm is stronger than anticipated, disaster can strike. Tall cranes can be particularly hazardous in stormy weather and may cause significant damage if they are blown over —as happened on this construction site in China.

THE WEATHER IS CHANGING...

... and so is the climate. It is happening very fast, it is happening everywhere, and it is largely down to us. Human activities have triggered a series of changes that will greatly affect all of our lives, all of our planet, and all of its wildlife. To lessen these effects, we must act. There is simply no time to lose.

Weather and climate do not stop at a country's borders. They are truly global and international. The problem of human-induced climate change is affecting the whole world. Of course, Earth's climates have changed in the past. There have been ice ages, hotter periods, and spells of floods and droughts. But modern climate change is different. It is taking place not over millions, or even thousands, of years, but in a few decades. The main causes are polluting gases being released into the atmosphere by our modern way of life. They are collecting in the atmosphere and making temperatures increase—a process known as global warming.

ROAD TO NOWHERE

Ever since the motor car was invented in the 1880s, traffic has increased year after year. So have traffic jams, backups, and the resulting air pollution that is a major factor in climate change. In the 1970s there were fewer than 400 million motor vehicles. Today there are almost 1.5 billion. One-fifth of these are in North America and a similar number in China.

A shroud of pollution envelops the city of Beijing, China, on a December morning as commuters travel to work.

THE BIG MELT

One of the most obvious signs of global warming and climate change is the melting of large bodies of ice, as seen here at Jökulsárlón Glacier Lagoon in Iceland. The lagoon has quadrupled its size since the 1970s as the glacier has melted and retreated. Melting is happening all around the world, from polar ice caps and glaciers to high mountains where the once permanently frozen soil is now turning to slush. The speed of melting is getting faster.

CLIMATES LONG AGO

Earth's climate has changed through history. Long before humans were around to affect things, temperatures, rainfall, and climate zones altered over time. The causes included tiny changes in the shape and distance of Earth's orbit around the Sun, and the angle of Earth's tilt (see page 40). These small shifts affected the amount of heat and other kinds of energy that reached various parts of our world. Evidence of our planet's past climates comes from analyzing many different things: minerals in rocks; tiny bubbles of air trapped deep in ice, rocks, and groundwater; and the preserved remains of plants and animals.

LIFE DURING ICE AGES

In the past 500 million years there have been about six or seven very cold periods called ice ages. The most recent ice age began around 110,000 years ago (see page 38). Ice sheets and glaciers spread from the poles toward the Equator, particularly in the Northern Hemisphere. The ice in the north covered its greatest area, known as the Last Glacial Maximum, around 25,000–20,000 years ago. Humans coped by avoiding the coldest areas, wearing warm animal furs, living in caves or shelters, and lighting fires for warmth and cooking.

SNOWBALL EARTH

Some scientists believe that extreme climate change caused the whole world to freeze from about 700–650 million years ago. This is known as the Cryogenian Ice Age, or Snowball Earth. It is possible that the seas and oceans around the Equator did not quite freeze over at this time, leading to an alternative name: Slushball Earth. However, the reasons for the deep freeze are not clear and other scientists dispute that these events even happened.

An artist's impression of a group of prehistoric hunters, dressed in animal furs, setting out in search of food

MOVING CLIMATE ZONES

During major ice ages, the planet's climate zones shifted and squeezed toward the Equator, as ice spread out from the poles. Patterns of rainfall also changed, meaning some regions switched from wet to dry, while dry places experienced more rain and floods. Scientific tests on samples of ancient rocks and minerals have shown that about 30,000 years ago, just before the last ice age ended, the temperature at the Equator was 9°F (5°C) lower than it is today.

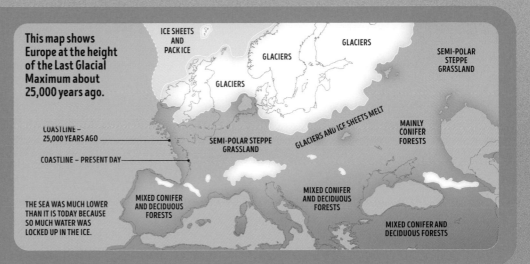

This map shows Europe at the height of the Last Glacial Maximum about 25,000 years ago.

ICE SHEETS AND PACK ICE

GLACIERS

GLACIERS

GLACIERS

SEMI-POLAR STEPPE GRASSLAND

GLACIERS

COASTLINE – 25,000 YEARS AGO

COASTLINE – PRESENT DAY

SEMI-POLAR STEPPE GRASSLAND

GLACIERS AND ICE SHEETS MELT

MAINLY CONIFER FORESTS

THE SEA WAS MUCH LOWER THAN IT IS TODAY BECAUSE SO MUCH WATER WAS LOCKED UP IN THE ICE.

MIXED CONIFER AND DECIDUOUS FORESTS

MIXED CONIFER AND DECIDUOUS FORESTS

MIXED CONIFER AND DECIDUOUS FORESTS

ADAPTED TO THE COLD

Evidence of previous climates is sometimes found in the thawing ice and soil of northern lands such as Siberia in Russia. This 39,000-year-old woolly mammoth was discovered there in 2010 and is in almost perfect condition. It was well adapted to freezing temperatures, with a very long, thick coat of hair to keep in body heat—strong evidence that the climate in Siberia at that time was extremely cold. As the ice extended out from the North Pole, animals and plants that were not adapted to the cold would have spread south to warmer places. When the ice fronts finally melted and retreated, the plants and animals moved their ranges north again, and many of the ice-adapted animals, such as mammoths and woolly rhinos, died out.

NOT SO GRADUAL

A few times in the distant past, Earth and its climate have undergone sudden, colossal change. Perhaps the most famous time was 66 million years ago, at the end of the Cretaceous Period. An asteroid 6 miles (10 km) wide smashed into the planet just off the coast of what is now Mexico. Gigantic clouds of ash, fumes, and dust blasted into the atmosphere, dimming the Sun's light and warmth for years. Ground tremors set off earthquakes and volcanic eruptions that poured out gases. Over three-quarters of all living things, including the dinosaurs, died out. Global climate patterns changed dramatically, and the evolution of life took new directions.

EARTH'S TEMPERATURE HISTORY

Scientists have used many kinds of evidence to estimate average world temperatures throughout prehistory. The story that emerges is one of ups and downs. One of the warmest times was during the middle and end of the Permian Period, around 250–230 million years ago. The world map was very different then, with all landmasses joined into one supercontinent, known as Pangaea, and one huge ocean, Panthalassa. Ocean currents, wind, rainfall, clouds, and other aspects of climate were very different from today, too. The Permian Period ended 252 million years ago with the Great Dying—the greatest mass extinction the world has ever known. Up to nine-tenths of all living things died out. The main cause was massive volcanic eruptions in what is now Siberia, northern Asia, which poured vast amounts of greenhouse gases into the atmosphere. This triggered a rise in global land and sea temperatures, reduced oxygen in the air and water, and made oceans more acidic. (Similar processes are at work today; see page 158.)

GLOBAL TEMPERATURE OVER LAST 500 MILLION YEARS

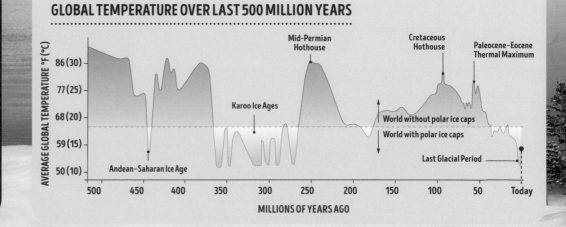

AVERAGE GLOBAL TEMPERATURE °F (°C)

86 (30)
77 (25)
68 (20)
59 (15)
50 (10)

Mid-Permian Hothouse

Cretaceous Hothouse

Paleocene–Eocene Thermal Maximum

Karoo Ice Ages

World without polar ice caps

World with polar ice caps

Andean–Saharan Ice Age

Last Glacial Period

500   450   400   350   300   250   200   150   100   50   Today

MILLIONS OF YEARS AGO

THE RISE OF INDUSTRY

The modern era of climate change is usually traced back to the beginning of the Industrial Revolution. In the late 1700s, newly invented machines took over laborious, repetitive work that was previously done by thousands of people. The first machines were powered by wind, water, and other natural forces, which limited where they could be located. Then engines were invented that were powered by the heat from burning fossil fuels—wood, coal, gas, petroleum, and diesel. Now machines could be located almost anywhere and grouped together in large numbers—inside a factory, for example. By the mid-to-late 1800s, giant factories and industries were springing up around the world—and Earth's climate was already changing.

UP AND UP

One of the main gases that causes global warming is carbon dioxide, CO_2 (see page 156). It is released when substances burn, which can happen in all sorts of different ways, from a simple wood fire to the latest gas-fired power station. The amount of CO_2 in the atmosphere is measured in parts per million, or ppm. By examining the levels of carbon stored in many natural objects, including rocks, trees, and glaciers, scientists have been able to figure out how CO_2 levels have risen over the past few centuries. Before the Industrial Revolution, levels around the world were about 280 ppm. By about 1800, this had risen noticeably, and it has kept on rising. Today, the figure is heading above 420 ppm.

⊙ Smoke pours out of the chimney of a coal-powered power plant. Coal-fired power stations are one of the largest sources of greenhouse gases in the world today.

CHANGES IN CO₂ LEVELS

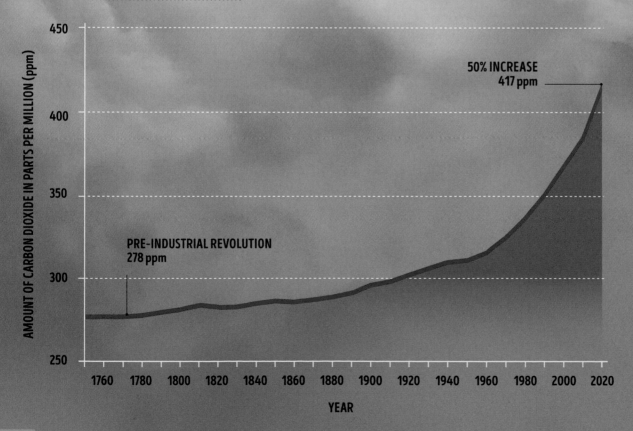

AMOUNT OF CARBON DIOXIDE IN PARTS PER MILLION (ppm)

450

50% INCREASE
417 ppm

400

350

PRE-INDUSTRIAL REVOLUTION
278 ppm

300

250

1760 1780 1800 1820 1840 1860 1880 1900 1920 1940 1960 1980 2000 2020

YEAR

INDUSTRIAL REVOLUTION: BENEFITS

The owners of coal mines, quarries, and factories made massive fortunes from industrialization. As more railroads opened, like the one illustrated below in Germany in 1839, long-distance travel became possible for those who had never previously been farther than a few miles from their home. All kinds of products, from essentials like food and clothing to luxury goods such as perfumes and silks, became more widely available and began to cost much less. However, this was all coming at the cost of climate change.

INDUSTRIAL REVOLUTION: DRAWBACKS

Rich factory owners made people work long hours for little pay. There were few laws to protect workers' health and safety, and injuries and deaths at work were common. In crowded factories, diseases spread fast. Burning wood, coal, and other fuels released choking smoke and fumes that covered whole towns and cities —and caused widespread breathing difficulties.

A 19th-century artist's impression of the industrial skyline of the West Midlands, UK

THE SPREAD OF PEOPLE

In 1800, the world's human population was about 1 billion and rising relatively slowly. But it soon began to speed up. By the year 2000, the human population was more than 6 billion. It is now approaching 8 billion. All these extra people need food and places to live. Around the globe, natural areas have been devastated by human use. This etching shows wetlands in Indonesia being drained in 1890 for a road and farmland. Loss of vast areas of plants, especially trees—known as deforestation —is one of the main causes of climate change (see page 176).

THE GREENHOUSE EFFECT

Almost since planet Earth formed 4.54 billion years ago, it has had a "greenhouse effect." The layer of air around the world—the atmosphere—works like the glass panels of a greenhouse to keep in, or trap, some of the heat and other energy from the Sun. Certain gases in the atmosphere have a more powerful greenhouse effect than others. The main ones are carbon dioxide (CO_2), methane (CH_4), nitrous oxide (N_2O), hydrofluorocarbons (HFCs), and water vapor (H_2O)—see pages 158–159. When the amount of these gases in the atmosphere goes up or down, the greenhouse effect increases or decreases, causing temperatures to change, as these five examples show.

1 EARTH:

NO GREENHOUSE EFFECT

If Earth had no atmosphere, and so no greenhouse effect, conditions would be very, very different. The surface temperature would be about 59°F (33°C) cooler.

EARTH TEMPERATURE	WITH greenhouse effect	WITHOUT greenhouse effect
Average	59°F (15°C)	almost 0°F (-18°C)
Warmest averages	113°F (45°C)	257°F (125°C)
Coldest averages	-58°F (-50°C)	-274°F (-170°C)

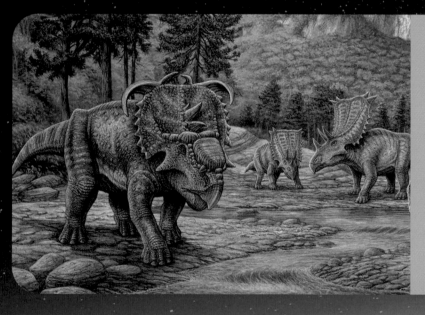

2 EARTH:

PREHISTORIC GREENHOUSE EFFECT

The Cretaceous Thermal Maximum, or the Cretaceous Hot Greenhouse, occurred 95–90 million years ago. Average temperatures were 13–14°F (7–8°C) above those of today, and there was probably no ice at the poles. Large reptiles such as crocodiles and chasmosaurs (left) lived in lakes and swamps in the Arctic Circle. One reason for this "greenhouse" event was raised carbon dioxide levels in the atmosphere due to many volcanic eruptions. However, recent research suggests there may have been other causes, too.

◎ The atmosphere wraps like a blanket around our planet and there are tiny traces of it even 6,200 miles (10,000 km) above the surface.

3 EARTH:

GLASS GREENHOUSE EFFECT

The Sun's light rays pass easily through glass to the inside of a greenhouse, but the infrared radiation, or heat, does not. Inside, objects and surfaces absorb the light, which has very short waves, then convert it into the longer waves of heat, which are then re-radiated back out into the air. However, like heat from outside, this inside heat cannot now easily pass through the glass. So it builds up inside the greenhouse. The weather may be below freezing outside, but inside it is warm enough for tropical plants to thrive.

Tropical plants flourish in winter in this UK greenhouse.

4 EARTH:

TODAY'S NATURAL GREENHOUSE EFFECT

Known collectively as solar radiation, many different types of energy —including light, heat, and ultraviolet—arrive on Earth from the Sun. Some of these are absorbed, or soaked up, by the atmosphere, land, and water, which get warmer. Some of the absorbed non-heat energy is changed into heat, too, especially by the land and by certain gases in the atmosphere. This "second-hand heat" is known as re-radiation and it adds to the warmth received directly from the Sun. Gradually, all these forms of energy come to a balance and keep the Earth at an average temperature, which today is around 59°F (15°C).

SOLAR RAYS

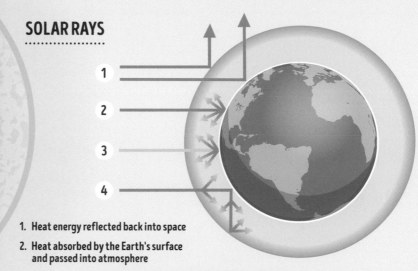

1. Heat energy reflected back into space
2. Heat absorbed by the Earth's surface and passed into atmosphere
3. Light and other non-heat energy absorbed and radiated as heat
4. Heat energy gradually absorbed by greenhouse gases in the atmosphere

5 NOT EARTH:

RUNAWAY GREENHOUSE EFFECT

Venus is the second planet from the Sun. Its atmosphere is almost all carbon dioxide (CO_2), a powerful greenhouse gas. This produces a massive greenhouse effect. The average Venusian temperature at the surface is 880°F (470°C). This is about the same as a wood fire that is starting to make flames and glow, and is hot enough to melt the metal lead.

GLOBAL WARMING IS HERE

Detailed calculations by scientists around the world show that our planet's temperatures are rising. This is due to increasing amounts of greenhouse gases being released into the atmosphere. With this global warming come changing climate patterns (see page 160). These will affect everyone, everywhere. Most of the people around the world understand the problem and are willing to take action, but it is much more difficult to figure out what these actions should be (see pages 176–177).

NUMBER ONE CULPRIT: BURNING FOSSIL FUELS

Burning fossil fuels is by far the greatest source of greenhouse gases, especially carbon dioxide (CO_2). The main fossil fuels are coal, oil, and natural gas. They were made over millions of years from the preserved remains of once-living things.

➤ Coal was once great forests of trees and other plants, which grew in warm, wet swamps about 300 million years ago (right).

➤ Oil came mainly from tiny living things in the oceans that sank to the seabed and began to decay. Petroleum, diesel, and similar fuels come from oil.

➤ Natural gas was also produced from rotting, decomposing plants and animals between 350 and 50 million years ago.

WHY CO_2?

Fossil fuels contain mainly hydrogen and carbon, so they are called hydrocarbons (HCs). When they burn, they combine with oxygen in the air to form the greenhouse gas carbon dioxide, CO_2, and also water vapor, H_2O. In chemistry this is written:

$$2HC + O_2 \blacktriangleright CO_2 + H_2O$$

This is the reverse of the plant process known as photosynthesis, which uses carbon dioxide and makes oxygen (see page 176). That is why fewer trees means more CO_2, and more plants and trees means less CO_2.

GLOBAL WARMING POTENTIAL (GWP)

The GWP measurement (see next page) compares the effects of different gases on global warming. Scientists have calculated the contribution to global warming made by 1 ton (1 tonne) of carbon dioxide added to the atmosphere. This has been given the value of 1 unit of GWP. Similar calculations for other gases are then made and compared to CO_2. For example:

1 ton (1 tonne) of methane has a GWP of 25–35.

This means that it has a global warming potential that is 25–35 times more than CO_2.

⊙ The cooling towers of a power station in Yorkshire, UK, trail clouds of water vapor—the most abundant greenhouse gas in the Earth's atmosphere—into the sky.

KEY GREENHOUSE GASES

The box on the right provides information on the main greenhouse gases that have been produced by human activity in the industrial era. The figures don't include gases from natural sources, such as wild animals and plants giving out carbon dioxide when they breathe, or respire. The "total emissions" figure is the amount there is of each gas in the atmosphere, shown as a percentage. However, it's not just the amounts of a gas that is important. Some gases have a much stronger greenhouse effect than others. Another important factor is how long these gases stay in the atmosphere—those that persist longer have a greater overall effect. Combining these factors gives a gas's "Global Warming Potential" score, estimated for a time period of 100 years.

GREENHOUSE GAS PRODUCERS BY ECONOMIC SECTOR

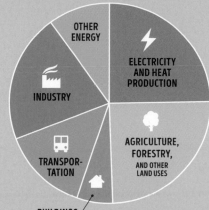

OTHER ENERGY

ELECTRICITY AND HEAT PRODUCTION

INDUSTRY

AGRICULTURE, FORESTRY, AND OTHER LAND USES

TRANSPORTATION

BUILDINGS

BREAKDOWN OF GREENHOUSE GASES IN THE ATMOSPHERE

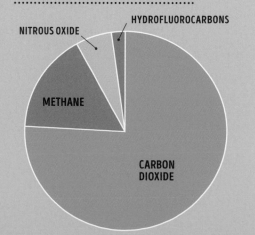

NITROUS OXIDE

HYDROFLUOROCARBONS

METHANE

CARBON DIOXIDE

*Water vapor not included since it is natural and varies greatly

CARBON DIOXIDE (CO_2)

Total emissions: 76%

Global Warming Potential (compared to CO_2): 1

Sources: Burning fossil fuels for warmth, for vehicles and transportation, and for electricity in power stations; also cutting down trees and removing other plants; major industry, such as cement production and general construction.

METHANE (CH_4)

Total emissions: 16%

Global Warming Potential (compared to CO_2): 25–35

Sources: Agriculture, including crops and farm animal emissions (intestinal gases); waste disposal and landfill; energy use; biomass burning.

NITROUS OXIDE (N_2O)

Total emissions: 6%

Global Warming Potential (compared to CO_2): 260–300

Sources: Agriculture, including making and using fertilizers; burning fossil fuels.

HYDROFLUOROCARBONS (HFCs)

Total emissions: 2%

Global Warming Potential (compared to CO_2): 150–20,000+

Sources: Industry, including making materials such as plastics; manufacturing products; refrigeration.

WATER VAPOR (H_2O)

Total emissions: Variable

Global Warming Potential (compared to CO_2): Negligible

Sources: Rain, oceans, rivers, living things. It's a strong greenhouse gas but, at present, the one that we can control least.

TOP 10 BIGGEST

GREENHOUSE GAS EMISSIONS – PER COUNTRY

These are the countries that emit most greenhouse gases. A major source is burning fossil fuels for industry.

		% WORLD TOTAL
1	CHINA	23
2	USA	11
3	INDIA	7
4	EUROPEAN UNION	7
5	RUSSIA	4
6	INDONESIA	3
7	BRAZIL	3
8	JAPAN	2
9	IRAN	1
10	CANADA	1

TOP 10 BIGGEST

GREENHOUSE GAS EMISSIONS – PER PERSON

These are the countries that emit most greenhouse gases per person. Some of them are quite small nations but they burn a lot of fossil fuels—for example, for heating in cold climates or air-conditioning in hot climates.

		TONS (TONNES)
1	AUSTRALIA	23 (21)
2	USA	22 (20)
3	CANADA	21 (19)
4	LUXEMBOURG	18 (17)
5	NEW ZEALAND	18 (17)
6	ICELAND	15 (14)
7	IRELAND	13 (12)
8	CZECH REPUBLIC	12 (11)
9	ESTONIA	12 (11)
10	THE NETHERLANDS	11 (10)

CLIMATES IN THE FUTURE

Earth's climates are changing, but what will they look like in the future? Global warming does not mean that our climates will be warmer everywhere. Some regions will actually be cooler. In fact, temperatures, clouds, wind, rainfall, and many other characteristics will all be affected. Regular reliable weather in some places could become more changeable, while unsettled weather in some places could become more unchangeable. Wet places may dry out and dry regions become wetter. Overall, extreme weather events are likely to be much more common.

PREDICTED CHANGES

There are many predictions for how climates might be different in years to come. Like our day-to-day weather forecasts, some will turn out to be more accurate than others. However, most of them agree on a few general trends. Global average temperatures and sea levels will rise. Today's climate zones (see page 42) will probably spread from the Equator toward the poles, so that equatorial and tropical climate zones widen and subpolar and polar zones shrink.

TEMPERATURES AND ALBEDO

At high latitudes, especially in the far north, temperatures are warming more than twice as fast as in places closer to the Equator. A major factor in these differences is albedo—the amount of the Sun's light and heat energy that is bounced back, or reflected, from a surface. Ice and snow have high albedos, reflecting back nearly all of the Sun's light and much of its heat, too. This helps them to stay cold. The problem is that global warming is melting ice and snow to expose darker soil, rocks, and other surfaces. These have lower albedos and so absorb more heat, speeding up the warming process.

OCEAN CURRENTS

The research vessel in the picture above is collecting data about ocean currents in the North Pacific. Oceans soak up heat and release it more slowly than the land. This means that there may be a delay of tens, hundreds, or even thousands of years before ocean currents react to changing temperatures and shift their direction.

RAINFALL

Most climate prediction systems show a future increase in rain, snow, and other precipitation around the world. However, much of this will fall toward the north and south. In the midlatitudes and Tropics, less rain is likely. This will endanger the last remaining areas of Earth's once-great tropical rainforests. Even though the overall amount of rain will lessen, it may well come in sudden huge bursts, like this flash flood in Germany in 2019, with longer dry periods in between, leading to cycles of floods and droughts.

📷 Rising temperatures and reduced humidity levels have increased droughts in some areas of the Namib Desert in Southern Africa.

MOUNTAIN CLIMATES

The warming atmosphere is already causing glaciers and snowfields to melt at lower altitudes. In the Scottish Highlands (above), scientists track the rising snowline to help measure the effects of climate change on the region.

TROUBLE IN STORE

In the Arctic tundra lands of the far north, shown here in Greenland, the soil should stay frozen, apart from a brief thaw during the short summer. But global warming has already affected millions of acres of this permafrost. In these areas, it has melted the ice in the soil, turning the ground into thick, muddy slush. Permafrost layers contain trapped gases, including greenhouse gases such as methane, carbon dioxide, and nitrous oxide. After being locked away in the ice for thousands, or even millions, of years, the thaw is now releasing them.

GLOBAL WINDS AND JET STREAMS

The high wind patterns that caused these cirrus clouds could change in the future. The atmosphere above lands and oceans will change, affecting the way winds blow and circulate around the world, and how the jet streams curve and wander. In turn, this will alter the way atmospheric heat moves around. Some places will lose their warming winds, others will gain them.

OCEANIC CLIMATES

Warmer oceans, rising sea levels, and more varied wind patterns make coastal climates very difficult to predict. Some could become more constant through the year, as the distinct seasons merge into one long, moderate season, with rainfall in most months. Other places could experience more severe storms, like this one off the coast of England in 2018.

CONTINENTAL CLIMATES

The centers of big continents are already dry and hot in summer and cold in winter. Global warming could cause these climates to become even more extreme, leading to larger areas of desert with extra-hot summers, slightly less severe winters, and longer droughts.

HURRICANES AND TYPHOONS

These massive tropical and subtropical storms gain much of their energy from warm oceans and have enormous destructive power, as seen here in the Philippines in November 2019. Like many extreme weather events, they are likely to become more common and more powerful in years to come.

GREAT BARRIER REEF: PARADISE IN PERIL

A location's biodiversity refers to the range and number of different organisms that live there. Among our planet's most biodiverse habitats are coral reefs—and the biggest of all is the world-famous Great Barrier Reef. It stretches along the northeast coast of Australia for about 1,400 miles (2,300 km) and varies in width from 37–150 miles (60–240 km). Made up of more than 3,000 smaller reefs and almost 1,000 islands, its vast range of corals, fish, turtles, and other sea life is unrivaled by any other habitat on Earth, apart from tropical rainforests. But the Great Barrier Reef is already suffering badly from the effects of global warming, and the damage is getting worse.

LOCATION

CORAL BUILDERS

As these images show, corals are actually formed of colonies of tiny creatures called polyps, which look a bit like miniature sea anemones. They, in turn, are home to microscopic life forms known as zooxanthellae algae. These algae capture energy from sunlight to make food, which they share with their coral hosts in return for a place to live.

UNDER ATTACK

Corals and their zooxanthellae are very sensitive to water temperature and acidity. As the oceans warm and more carbon dioxide dissolves in them—making them more acidic—the corals and their zooxanthellae begin to die. This leaves behind only the polyps' bleached white skeletons, and the reef becomes a bare, deserted graveyard—as this picture, taken during a bleaching event on the Great Barrier Reef in 2001, shows.

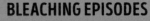

BLEACHING EPISODES

Some episodes of coral bleaching are temporary. The coral polyps get rid of their zooxanthellae partners and shut down into survival mode. Their colors fade but they are not actually dead (see below). If conditions improve, the polyps may revive. More and more often, however, seawater heat waves are "cooking" the Great Barrier Reef and fewer corals recover. Mass bleaching episodes are becoming more frequent, more widespread, and more serious.

UNSPOILED REEF

Parts of the Great Barrier Reef (above) are still healthy. They offer an amazing display of vibrant colors, with corals, fish, worms, octopuses, turtles, and hundreds of other species all living together in one of Earth's most complex habitats. But surveys show that between 1985 and 2020, the whole reef lost about half its corals. The species that build large, spectacular, branching and flat-topped structures were the worst affected.

MULTIPLE THREATS

Apart from global warming, the Great Barrier Reef faces many other threats:

⚠ Rivers draining into the sea carry chemicals used in farming, such as fertilizers and pesticides, which harm the sea life.

⚠ Rivers also carry silt, mud, and other polluting particles that cloud the water and choke the corals.

⚠ The reef has a population of big, spiny, venomous crown-of-thorns starfish, which eat the coral. Every so often, the starfish multiply to plague-like numbers and devour large areas of coral.

⚠ The changing El Niño–La Niña cycle of the Pacific Ocean (see pages 24–25).

⚠ The increasing incidence of severe weather episodes, such as storms, which can damage the reef's structures.

MORE EXTREME EXTREMES

Most climate change computer models predict more extreme weather events (see pages 80–113). Many suggest that even more frequent weather extremes will occur in places that already endure them. They also warn of serious weather events occurring in places where they were previously rare or unknown. One of the main factors in these predictions is that much of our planet's weather—especially high winds, storms, heavy rains, and extensive floods—are powered by heat from the Sun. Global warming means more of this heat, resulting in more frequent and more widespread episodes.

HURRICANES AND OTHER STORMS

In 2021, a major scientific report made the following predictions about hurricanes for the rest of the 21st century:

⚠ A higher proportion of all hurricanes and similar tropical storms that form will reach the strongest levels of Categories 4 and 5 (see page 109).

⚠ Of these Category 4 and 5 storms, more will become even stronger than they currently are.

⚠ The rainfall from all strengths of hurricanes and similar storms will increase by as much as 15 percent.

⚠ Coupled with rising sea levels, these bigger, more powerful storms will flood much greater areas of land than at present.

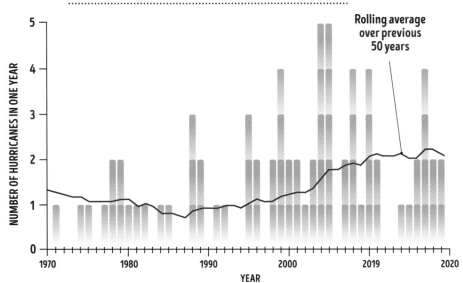

CATEGORY 4 AND 5 ATLANTIC HURRICANES PER YEAR

Rolling average over previous 50 years

NUMBER OF HURRICANES IN ONE YEAR

YEAR

DROUGHTS

From massive factories to houses and homes, modern life uses lots of water—and water use is predicted to rise, leading to what scientists call water stress. Some dry regions are forecast to receive even less rain because of global warming. The Al Massira dam and reservoir in Morocco, North Africa (above), is a vital water source for local people and farming. However, the reservoir has shrunk significantly, especially from 2015, with water loss around its shallow edges.

FLOODS

Climate change will probably contribute to greater flooding in three ways. Firstly, sea levels are rising, so seawater will be more likely to flood onto low-lying coasts (see page 166). Secondly, increasingly powerful storms will create higher waves and bigger storm surges onto shorelines. Finally, more rain will raise the levels of rivers and lakes so they overflow their banks onto the surrounding land. A recent study on these risks for the coastal city of Charleston, South Carolina (below), suggested that the number of tidal floods could increase from fewer than 20 per year in the early 2000s to more than 150 by 2050.

STORM OF THE CENTURY

March 1993 saw the Storm of the Century (below) move north from the Gulf of Mexico, Central America, and the Caribbean across the USA toward Canada. Its combination of gigantic size, power, widespread snowfall, winds, and floods had never been recorded before. The nickname Storm of the Century gives a good idea of how rare this event was, and it is still remembered by this name today.

HOW OFTEN IS A "ONCE IN 20 YEARS" EVENT?

Extreme episodes are sometimes called "once in a decade" or "once in a century" events, but what does that mean? In climatology, this information can be expressed as the "recurrence interval," or RI, and the annual exceedance probability, or AEP. Imagine that weather records for a location show that a certain figure for annual rainfall was exceeded five times over the past 100 years. This does not mean there will definitely be such an event every 20 years. It shows, on average, the RI is 20 years. In any individual year, the AEP for excess rain is 1 in 20, or 5 percent. After years of low rainfall, there could be several heavy rainfall years in a row, then many more years of low rainfall. All of these statistics, analyses, and predictions can only give an estimate of what will happen. What is generally agreed, though, is that global warming is making "big weather" more common.

📷 Snowplows roll down First Avenue in New York City trying to keep on top of the blizzard conditions of March 1993.

RISING SEA LEVELS

Climate change is causing sea levels to rise for two main reasons. Firstly, like most substances, water becomes larger as it is heated—known as thermal expansion—so warming oceans are increasing in volume. Secondly, global warming is melting ice caps, ice sheets, and glaciers, and much of this meltwater, even inland, eventually finds its way into seas and oceans. However, sea level rise is complicated, and it is not happening everywhere at the same rate. For example, some northern landmasses were squashed down by the incredible weight of glaciers during the last ice age. With that weight gone, they are slowly rising up again—faster than sea levels are increasing. In these places, sea levels actually appear to be falling.

DIRE PREDICTIONS

If all the world's ice melted, sea levels would rise by about 230 ft. (70 m). Every coastal city on Earth would be deep under water, as would many inland areas. Hopefully, this will never happen. Even so, the threats from rising sea levels are massive. About one-tenth of the world's population live on coasts and inland areas that are less than 33 ft. (10 m) above present sea levels. Most current forecasts predict a rise of up to 20 in. (0.5 m) by 2100 unless the world takes drastic action to slow global warming.

CITIES MOST THREATENED BY RISING SEA LEVELS

Percentage of cities' population affected by rising sea levels with a 7.2°F (4°C) rise in global temperature

+7°F

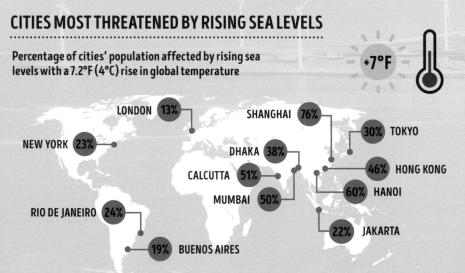

- LONDON 13%
- SHANGHAI 76%
- TOKYO 30%
- NEW YORK 23%
- DHAKA 38%
- HONG KONG 46%
- CALCUTTA 51%
- HANOI 60%
- MUMBAI 50%
- RIO DE JANEIRO 24%
- JAKARTA 22%
- BUENOS AIRES 19%

PREDICTIONS FOR RISES IN SEA LEVEL

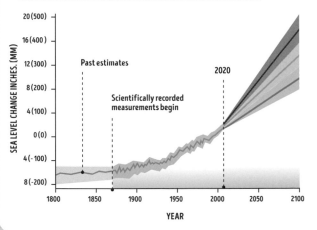

SEA LEVEL CHANGE INCHES. (MM)

- 20 (500)
- 16 (400)
- 12 (300)
- 8 (200)
- 4 (100)
- 0 (0)
- 4 (-100)
- 8 (-200)

Past estimates

Scientifically recorded measurements begin

2020

YEAR: 1800, 1850, 1900, 1950, 2000, 2050, 2100

SCENARIO 1: Little change in current levels of industrial growth and carbon emissions.

SCENARIO 2: Moderate change in reducing carbon emissions and in changing to "green" industrial processes using renewable energy.

SCENARIO 3: Drastic reduction in carbon emissions and almost all industry becomes "green" and sustainable.

BATTLE AGAINST THE SEA

The Netherlands is one of the world's lowest-lying countries. Its average height is only 100 ft. (30 m) above sea level. The Delta Works (shown here) is a mammoth barrier project designed to keep out the sea, mainly during high tides and storm surges. The first stage was completed in 1997. Now that the risks are clearer about rising sea levels due to global warming, plans are in place to make it even higher and stronger, at a gigantic cost perhaps exceeding $100 billion.

ALREADY GOING

The Marshall Islands, near the Equator in the western Pacific Ocean (below), is one of the world's lowest-lying countries. Its average height across all its islands and islets is less than 5 ft. (1.5 m) above sea level. Already, the ocean is flooding in more than it did. If levels rise by just 18 in. (500 mm) more, almost half of the inhabited parts of the islands will be submerged.

ONE-TENTH SMALLER

About two-thirds of Bangladesh is less than 16 ft. (5 m) above sea level, including huge cities such the capital city Dhaka (below). Even a rise of 20 in. (500 mm) would flood more than one-tenth of the land, including huge seaports vital for trade, large areas of farmland, and the homes, businesses, and industries of more than 20 million people.

TIMELINE: CLIMATE CHANGE AND SEA LEVELS

⚠ **1838** French scientist Claude Pouillet suggested that if levels of carbon dioxide gas in the atmosphere rose, this could trap more heat and make the world warmer. It was one of the first proposals on the topic of climate change.

⚠ **1899** US geologist Thomas Chamberlin wrote: "It has been shown that the carbon dioxide and water vapor of the atmosphere have remarkable power of absorbing and temporarily retaining heat rays ... the effect is to blanket the Earth with a thermally absorbent envelope."

⚠ **Early 1900s** Tide gauges and sea level markers indicated that a rise in sea levels was beginning.

⚠ **1909** British scientist John Henry Poynting introduced the term "greenhouse effect."

⚠ **1968** The US Stanford Research Institute reported: "If the Earth's temperature increases significantly, a number of events might be expected to occur, including ... the melting of the Antarctic ice cap, a rise in sea levels, warming of the oceans...."

⚠ **1981** US climatologist James Hansen predicted: "Potential effects on climate in the 21st century include ... shifting of climatic zones, erosion of the West Antarctic ice sheet with a consequent worldwide rise in sea level."

⚠ **1992** The first satellites were launched into space to measure sea levels around the world and over time.

⚠ **2015** Five small islands belonging to the West Pacific Solomon Island group were lost due to rising seas and wave erosion from stronger winds and higher tides.

CLIMATE CHANGE AND NATURE

In the distant past, climate change usually happened very slowly, over thousands or millions of years. Plants and animals had time to evolve and adapt, or to gradually move and spread with their climate zones to new places. Climate change today is happening so rapidly that many kinds of habitats and wildlife are already in trouble—and when one living thing is affected, there is a knock-on effect on many others. A food chain is a sequence where a plant is eaten by an animal, which is consumed by a different animal, and so on. In this way, the lives of many living things are all connected. If climate change affects just one or two links, the whole chain is disturbed.

NATURE UNCOUPLED

A major challenge facing wildlife is the "uncoupling" of plants and animals in food chains and food webs.

⚠ Some plants follow a seasonal and yearly cycle based partly on daylight, as this is where their energy comes from. When the days get longer in spring and summer, this encourages plant leaves to unfurl and flowers to open.

⚠ Some animals, however, rely more on temperature. For example, butterflies and other insects lay their eggs as it becomes warmer, so their young have plenty of plant food.

⚠ Daylight cycles are not changing, but global warming is affecting temperatures around the world. If it becomes warmer earlier in the year, caterpillars hatch from their eggs earlier, but their leafy food may not yet have grown. In other food chains, a plant and the animals that eat it may both be affected by global warming, but at different rates.

⚠ This is leading to mismatches, or uncouplings, in natural cycles that have been synchronized for perhaps millions of years. For example, bees depend on rising temperatures to become active in spring. However, flowers, which rely on insects like bees for pollination, tend to be affected by daylight length. This could eventually lead to a serious mismatch between the timings of the flowers and pollinating insects.

THE EARLY BIRDS

To breed, some birds migrate long distances to places with good conditions and plentiful food. A recent scientific study looked at the timings of bird migrations on five continents using records dating back 300 years. The results showed that, on average, birds arrive at their breeding places about one day earlier for each 1.8°F (1°C) rise in temperature. This may not sound like much, but it could cause concerns. Local flowers, insects, and other foods may be changing their seasonal timings for food, and breeding at different rates from the migrants. So the migrants could arrive too early or too late to be able to utilize their usual habitats.

Barn swallows (above) need a plentiful supply of insects for themselves and their young. Climate change may affect the plants these insects feed on, and therefore the time when the insects are most numerous. Meanwhile, the swallows' migration dates are changing at a different rate, making mismatches between the birds and their food source ever more likely.

OUR FOODS AT RISK

Insects visit flowers to feed on their nectar. During these visits, they often also pick up pollen, which pollinates other plants. More than one-tenth of the world's food production, especially fruit farming (below), relies on insects pollinating crop flowers. Climate change experts predict how various kinds of mismatches could affect this process:

1 ### WRONG TIME, RIGHT PLACE

TEMPORAL MISMATCH: This is when the timing between most insects being active and most flowers making pollen uncouples, with one altering faster than the other.

2 ### RIGHT TIME, WRONG PLACE

SPATIAL MISMATCH: This happens when habitats change due to global warming. Plants with richer nectar may begin to appear in new environments, attracting insects away from our orchards and other plant and fruit crops.

3 ### RIGHT TIME AND PLACE, WRONG SIZE AND SHAPE

PHYSICAL (MORPHOLOGICAL) MISMATCH: Some species of insects tend to be smaller in warmer places, while certain flowers grow larger in higher temperatures. This could lead to a problem— the smaller insect's strength and nectar-gathering abilities may be unsuitable for bigger flowers.

4 ### RIGHT TIME AND PLACE, WRONG SIGNAL

RECOGNITION MISMATCH: To attract a specific kind of pollinating insect, some flowers give off a distinct scent. Raised temperatures can alter the scent flowers give off so that the insects no longer recognize it.

📷 A cherry orchard in the UK blooms in spring.

POSSIBLE WINNERS

There will probably be species that take advantage of global warming and climate change. Many living things are "mobile generalists." This means that they can move (or spread easily), cope with a range of habitats, and eat many different foods. Animals that could benefit include jellyfish, orcas (killer whales), frogs, snakes, gophers, and, worryingly, mosquitoes (below)—which already spread many killer diseases.

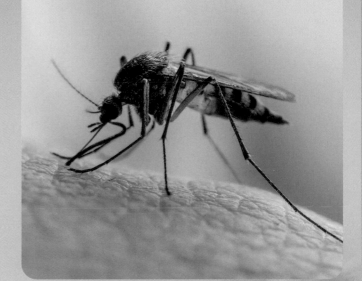

FUTURE SCENARIOS

No one knows precisely what the future holds or how climate change will progress. Scientists, such as those working with the Intergovernmental Panel on Climate Change (IPCC), use computer models to show what could happen. They create possible scenarios according to different conditions and show how they might alter depending on what we choose to do. Every few years, climate experts and government officials from around the world get together to discuss the latest set of IPCC scenarios to decide what their response should be.

THE IPCC

The Intergovernmental Panel on Climate Change is an international body that was set up in 1988 by the World Meteorological Organization (WMO) and the United Nations Environment Programme (UNEP). Its mission is to assess information relating to human-induced climate change. The IPCC represents almost every nation in the world. However, each country has its own interests. Some poorer countries do not have as good a standard of living as rich ones, and so want to develop their industries to try and catch up. Some wealthy nations, which can afford to reduce greenhouse emissions, may have trouble convincing their populations to make the changes to their lifestyles needed to cut greenhouse emissions.

SCENARIOS FOR GLOBAL WARMING

The IPCC studies more than 40 scenarios. It is a large and complex international effort involving all kinds of sciences. In simple terms, it shows what could happen if the world were to reduce its greenhouse gas emissions and industrial activity by a small amount, a medium amount, and a very large amount.

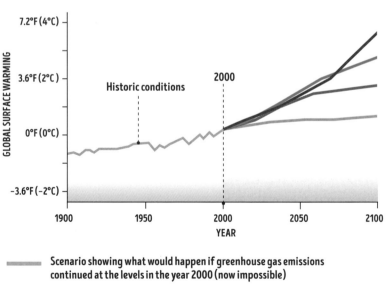

4 GLOBAL WARMING SCENARIOS

Scenario showing what would happen if greenhouse gas emissions continued at the levels in the year 2000 (now impossible)

Scenario in which greenhouse gas emissions reduce faster than planned in 2020

Scenario for greenhouse gas emissions that reduce as planned in 2020

Scenario if greenhouse gas emissions reduce more slowly than planned in 2020

A polar bear stands on the edge of a melting ice floe in the Arctic. Polar bears have become a symbol for the plight of creatures whose habitats are being destroyed by climate change.

BIG MEETINGS

The United Nations has been hosting an annual international climate change meeting since 1995. Organized by a UN group called the Conference of the Parties (COP), these meetings are officially known as UN Climate Change Conferences, and unofficially as COP meetings (the first was COP 1). They usually result in some kind of agreement, but these are often weakened by countries acting in their own interest. Many people believe that the COP meetings have been too slow in coming up with meaningful changes.

- **1997 JAPAN** The COP 3 Kyoto Protocol suggested that industrialized countries should reduce greenhouse gases according to targets set by each country for itself.
- **2009 DENMARK** COP 15 in Copenhagen agreed a "political intent," or aim, to limit carbon emissions and respond to climate change, both short and long term.
- **2015 FRANCE** COP 21 in Paris agreed to "pursue efforts" to keep global warming below 2.7°F (1.5°C) this century.
- **2021 SCOTLAND** At COP 26 in Glasgow, rich countries agreed to help poorer ones act against climate change.

At these kinds of meetings, many people act in their own country's interests. Others see progress as hopelessly slow.

TIPPING POINTS

The idea of climate tipping points has been around for a few decades. The idea is that small changes, especially if they affect each other, can suddenly lead to much larger ones—and these bigger changes may not be reversible. Scientists regard a global temperature rise of 3.6°F (2°C) above pre-industrial levels as a major tipping point. Serious effects, which you can read about in more detail on previous pages, include:

1 Melting of the West Antarctic and Greenland Ice Sheets

2 Changes in ocean currents such as the Pacific's El Niño–La Niña and the Atlantic Meridional Overturning Circulation (AMOC)

3 Further loss of the Amazon and other rainforests, which absorb and store carbon and produce oxygen

4 Thawing permafrost across far northern lands

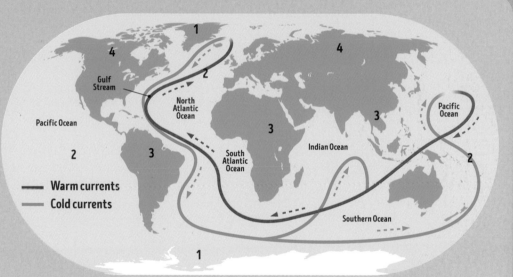

In the system of ocean currents known as the Atlantic Meridional Overturning Circulation (above), water cooled in the North Atlantic sinks and circulates at great depth all the way to the Indian and Pacific Oceans near the Equator. Here, it is warmed, rises to the surface, and returns. The whole cycle may take 1,000 years. Climate change could cause the cycle to weaken, affecting the movement of heat and energy around the globe.

A NEW AGE?

Geologists divide Earth's history of more than 4.5 billion years into eras, which are split into periods, which are then split into epochs. The current epoch, called the Holocene, began 11,650 years ago, at the end of the last major ice age. But climate change is now so serious that some scientists have suggested naming a new epoch—the Anthropocene (from *anthropo*, meaning "relating to humankind"). This would cover the period since the late 18th century when industry began to cause large-scale atmospheric changes. However, it is not yet agreed on as an official epoch.

CASE STUDY:

CHAMPION ECO-CITIES

As climate change progresses and people become more aware of this gigantic problem, some places are taking more action than others. "Eco-cities" is the name given to urban centers looking to develop in a clean, green, ecologically friendly way. This means trying to have energy-efficient buildings, setting aside green spaces, and promoting cycling and public transit. Industries and businesses are encouraged to use renewable energy and materials from sustainable sources, to create low levels of pollution, and to employ high levels of recycling. These forward-thinking cities can show the world a positive way forward for humankind and the planet.

LOCATION

COPENHAGEN, DENMARK

Public transit and cycling are very popular in the Danish capital Copenhagen (below and right). Fewer than one-third of families own a car. There are more bicycle-only lanes per person than in nearly any other city in the world. Local organic food, produced without artificial chemicals and fertilizers, makes up one-quarter of all food sold. Copenhagen's stated aim is to be carbon-neutral by 2025.

GANDHINAGAR, INDIA

Known as the greenest city in Asia, more than half the area of Gandhinagar (left) is covered with vegetation. The city is carefully planned so that most roads are lined with trees and parkland. Different kinds of buildings are organized into sectors—so industrial factories, business offices, shops and retail parks, and residential buildings are all in different zones. This means that busy freight railroads and roads that supply factories are not spread all around the city, taking up land and causing widespread pollution.

CURITIBA, BRAZIL

In Curitiba, the authorities run a scheme whereby residents can swap waste for food (right). People take their recyclables and other unwanted materials to a collection point where they are weighed and then exchanged for fresh food. In this *cambio verde*, or "green exchange," no money changes hands. Other things that can be swapped for waste include tickets for public transit and school books. The city now recycles more than three-quarters of its waste.

CLIMATE AND THE CARBON BUSINESS

Carbon is big business. To slow down climate change, limits on carbon emissions have been set both by international agreements and by laws passed in individual countries. These limits will become even lower in the future, putting pressure on countries to clean up their act even more. As a result, in the last few decades, a whole new area of business has grown up around buying, selling, swapping, and trading in carbon.

CARROT AND STICK

Many nations use a "carrot and stick" approach to reducing carbon emissions. The "carrot" encourages lower emissions by providing grants, loans, and tax breaks for new projects that are "low-carbon" —low in pollution and waste and efficient at recycling. The "stick" punishes those who break the rules, with fines, higher taxes, loss of trading licenses, and, in very serious cases, by putting people in prison. Old power stations and factories that use fossil fuels in an inefficient way must be especially careful of these changing rules.

📷 In the port of Nakhodka in Russia, coal is loaded onto a ship. Coal is up to four-fifths carbon and involves the most costly carbon taxes and tariffs.

CARBON TRADING

Carbon trading is when a company or industry buys permits or credits that allow it to produce a certain amount of greenhouse gas emissions. The hope is that by making companies pay high prices for their emissions they'll be encouraged to switch to cleaner, renewable sources of energy, such as solar power, to save money.

CAP-AND-TRADE

In some regions, companies can buy or sell credits between themselves. This is sometimes known as "cap-and-trade." The government or an international agreement sets a maximum amount for carbon emissions, known as the "cap." The "trade" part happens at an auction, where companies sell and buy their carbon allowances and credits. Like many markets, prices and amounts vary according to supply and demand. At the COP 26 climate talks (above) in Glasgow in 2021, nearly 200 countries agreed on rules for how international carbon markets should work.

CARBON CAPTURE AND STORAGE

In addition to reducing new carbon emissions from getting into the atmosphere, there are also ways to collect, or capture, carbon dioxide that is already there. This is known as carbon sequestration (see below). One method uses chemicals and filters to extract carbon dioxide gas from the atmosphere. It can also be used to prevent the carbon from being released in the first place—for example, by removing it from power station chimneys. The gas is then squashed, or pressurized, to make it liquid, and injected into porous, or sponge-like, layers of rock deep below ground. Another method, shown on page 178, uses the living world of plants and trees to capture the carbon.

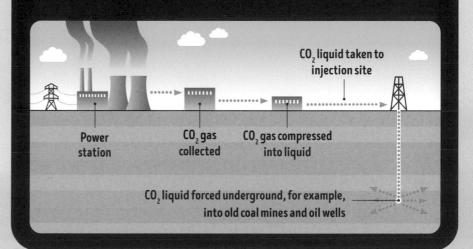

CO_2 liquid taken to injection site

Power station

CO_2 gas collected

CO_2 gas compressed into liquid

CO_2 liquid forced underground, for example, into old coal mines and oil wells

CARBON OFFSETS AND THE ETS

A "carbon offset" is an agreement to lower carbon emissions in one place thereby allowing them to stay the same—or even rise—in another. Like other forms of carbon trading, there are massive business markets for selling and buying offsets and similar deals. One of the biggest is the European Union's Emissions Trading System (EU ETS). It began in 2005 and represents thousands of power stations, factories, and industrial centers, such as the gigantic Volkswagen AG Wolfsburg car plant in Germany (below).

CARBON TARIFFS

A carbon tariff is an extra amount of money—a tax—added to the price of products that are "carbon-intensive." This means the raw materials, energy, and processes used to make them cause high carbon emissions. The tariff is usually added when the goods are imported, or brought into a country, such as this container seaport in Felixstowe, UK. It is designed to make carbon-intensive items more expensive, so fewer are sold. This should encourage the industries producing them to become less carbon-intensive.

SOLVING CLIMATE CHANGE

Our planet, its weather systems, its wildlife, and our own ways of life are in great danger. Climate change is speeding up—and to combat it, big businesses and industries around the world, including manufacturing, agriculture, food processing, and transportation, all need to make massive changes. Ordinary people can also make a big difference by signing petitions, engaging with websites and social media, sending letters and e-mails, joining climate change groups putting pressure on politicians, and generally getting the message out there. Here are some of the changes that could help put us back on the right course.

CONSERVE FORESTS

Forests are vital to our weather and climate. They absorb carbon dioxide, helping to counteract global warming—so we need to look after them. Deforestation is having a major impact on climate change. People can help by avoiding products that are being grown on cleared rainforest land or in a non-sustainable way.

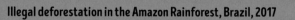

Illegal deforestation in the Amazon Rainforest, Brazil, 2017

CHANGE FARMING METHODS

Intensive "factory" farming using chemical fertilizers and pesticides, especially for meat production, is one of the top five carbon emitters. Raising cows, pigs, sheep, chickens, and other farm animals, especially in massive sheds with factory-made feeds, is a huge source of methane and carbon dioxide emissions. Growing plants for food is a much less polluting way of using farmland.

FEWER FOSSIL FUELS, MORE RENEWABLES

Mining for coal and drilling for oil and gas release large amounts of greenhouse gases, especially methane. Among the greatest carbon producers are fossil-fuel-burning energy companies, in particular, electricity suppliers. Consumers can help by supporting companies and industries that use more solar, wind, and hydroelectric energy.

BUY FEWER MANUFACTURED PRODUCTS

Fast fashion and cheap new clothes; changing smartphones and other gadgets when not necessary; buying mountains of disposable plastic toys and other consumer goods; getting a big, shiny, fossil-fuel-guzzling new car each year—all these harm the environment and contribute to global warming.

WASTE LESS ENERGY

Wasting energy, particularly electricity and heat, is a tremendous problem. From poorly insulated buildings to forgetting to turn off lights and appliances, we all waste huge amounts of energy every day. One positive change we can make in our homes is to switch from "dirty" coal, oil, or gas heaters and boilers to clean, green, efficient heat pumps.

CONSIDER TRANSPORTATION OPTIONS

Electric and hybrid vehicles (which use both gasoline and electric power) are better for the environment than ones that use only gasoline (or diesel) because they use less energy overall and produce less air pollution. We also need to question our travel choices. Is every single-person car trip really necessary? Walking, cycling, and using public transit are all effective ways to help the environment, save energy, and reduce pollution.

GRAMS OF CO$_2$ PER PASSENGER PER KM

Mode of transportation		Average no. passengers
Electric train		x 156
Car		x 4
Minivan		x 4
Electric bus, tram, coach		x 12.7
Scooter		x 1.2
Car		x 1.5
Minivan		x 1.5
Airplane		x 88

0   50   100   150   200   250   300

Many cities, such as São Paulo in Brazil, have fleets of "clean" electric buses.

📷 Vast open-cast mines, like this one in China, scar huge areas of land and it is very difficult and costly to restore the land to a natural environment.

CHANGES ARE COMING

The past few pages have shown some of the many ways we have already harmed our planet—and are continuing to do so. Temperatures are rising, glaciers are melting, coral reefs are under stress, and forests are disappearing. Storms, floods, and droughts are on the increase, and nature's habitats and food webs are in trouble. More severe weather and climate change will affect all of us, our ways of life, and indeed the future of the whole Earth. Yet we can all help—by changing our lifestyles, our decisions, and what we choose to do in the future.

NATURE IS KEY

One of the best ways to protect the climate is to get involved with nature. A key activity is to plant trees and other vegetation. As they live and grow, these plants take in carbon dioxide and give out oxygen—a process called photosynthesis. It is a double positive for the climate. Trees, bushes, and other plants also attract insects, birds, and other wildlife. Rewilding projects aim to restore damaged natural habitats.

Schoolchildren in Mozambique stand proudly by trees they planted in their playground.

DO MORE R-ING

Reduce, Repair, Re-use, Restore, Recover, Recycle —these should be our watchwords. Try to avoid single-use items and too much packaging. Consider making use of pre-loved items and things that have hardly been used, which are often given away for free. Recycle old fruit and vegetable peelings and green garden waste in a compost bin. Speak out and complain, for example, if a local organization has lots of single-use plastic, or if it seems to send lots of waste to the landfill, rather than recycling.

LOCAL, SEASONAL, HEALTHY

Think about what you eat. Meat production is a major producer of greenhouse emissions, while vegetables, fruits, and other plant foods are good for our bodies and for the climate. Also, check "food miles," which is how far an item has traveled from its field to your plate. Some food products are shipped halfway around the world so we can buy them whenever we want—but at great cost and damage to the climate.

VOICES OF THE FUTURE

Today's younger people will inherit the climate situation left by previous generations. Many of the most prominent climate campaigners are young. They speak out in the hope of affecting older adults who make laws, make political decisions, and run big businesses. But it can be difficult to make progress, as some of these older figures are interested in maintaining their wealth and power at the expense of meaningful action. Changing the attitudes of those in power is one of the greatest challenges of the climate change crisis.

Climate activist Greta Thunberg

CARBON FOOTPRINTS

Each person, process, event, and place has a carbon footprint. This is the amount of greenhouse gas emissions it causes. Many carbon footprint calculators are available online. Choose one and put in some imaginary answers that you think will produce a large carbon footprint, such as taking vacations abroad each year. Check the score. Now try again with answers that you think will leave a small carbon footprint. Find out which answers change the footprint value the most—then try to take those actions.

GLOSSARY

ANTICYCLONE Movement of air around a center of high pressure, usually bringing dry, settled weather. Anticyclones move clockwise in the Northern Hemisphere and counterclockwise in the Southern Hemisphere.

ATMOSPHERE The layer, or blanket, of air around Earth. It becomes thinner, or less dense, with altitude.

ATMOSPHERIC PRESSURE The pressing, or pushing, effect, due to the weight of air in Earth's atmosphere. Air does not just press down, but up, sideways and in all directions.

CARBON DIOXIDE (CO₂) A gas made by almost any kind of burning or combustion, and breathed out by many living things, including all animals. It is one of the main greenhouse gases that causes global warming.

CLIMATE The long-term weather and atmospheric conditions in a region, including temperature, rainfall, wind, and similar features. Climates are usually described over decades or hundreds of years; "weather" describes more short-term phenomena, over days and weeks.

CLIMATE CHANGE The altering, or shifting, of our planet's weather and climate patterns. Today it is mainly caused by human activity creating extra amounts of greenhouse gases in the atmosphere.

CLIMATOLOGIST A scientist who studies how the weather and climate work and how these change over decades and centuries.

CONVECTION CURRENT Circular movement of air, water, or other flowing substance, which usually carries its heat from a warmer to a cooler place.

CYCLONE Movement of air around a center of low pressure, usually bringing cool, wet, windy, unsettled conditions. Cyclones move clockwise in the Northern Hemisphere and counterclockwise in the Southern Hemisphere (not to be confused with a tropical cyclone; see right).

DEPRESSION An area of low atmospheric pressure, causing winds to move in a clockwise direction in the Northern Hemisphere and counterclockwise in the Southern Hemisphere. It usually brings wet, windy, cool, changeable weather.

DESERT A dry, or arid, region that receives an average of less than 10 in. (250 mm) of rain, snow, or other precipitation each year.

DESERTIFICATION When an area becomes like a desert, with only tough, drought-resistant plants and few animals. It often occurs when human activity ruins the original habitat, for example, by planting crops and introducing grazing animals.

DRAINAGE BASIN A region where all the rainfall runs into many streams, rivers, and lakes, which then drain into one particular river.

DROUGHT A long period of time when the average rainfall in an area is much less than its normal amount, causing a lack of water.

ELECTROMAGNETIC RADIATION Energy that travels through space in the form of waves, and includes radio waves, microwaves, infrared, visible light, ultraviolet, and X-rays.

EVAPORATION When a liquid turns into a gas, or vapor. In weather, it happens when water changes into water vapor, for example, when a puddle dries, and is usually caused by the Sun's heat.

FLOOD When water levels in streams, rivers, and lakes rise, caused by heavy rain. Eventually, the water has nowhere to go and spills out over the surrounding land.

FORECAST The prediction of what might happen to the weather or climate in the future. Weather forecasts tend to cover the near future, while climate forecasts may look long into the future.

FOSSIL FUEL A substance that can be burned to release heat energy that is made from fossils (the long-dead remains of plants, animals, and other living things). The main fossil fuels are coal, oil (petroleum), and natural gas.

GENERATOR A machine that makes electricity from another type of energy, such as heat, wind, waves and tides, solar (from the Sun), or nuclear (splitting the nuclei of atoms).

GLACIER A "river of ice" that slowly slides downhill due to gravity, usually in a valley.

GLOBAL WARMING The process that is causing the average temperature on our planet to gradually rise. It is due to the increase in the amount of greenhouse gases in Earth's atmosphere, which trap the Sun's heat.

GREENHOUSE GAS A gas in the atmosphere that traps the Sun's incoming heat. The main greenhouse gases caused by human activity are carbon dioxide (CO₂) and methane.

GULF STREAM A huge ocean current of moving water that carries warmth from the subtropical western Atlantic Ocean and Caribbean regions to Iceland and northwest Europe.

HIGH PRESSURE When the air is more dense and heavier, or "thicker," and so presses down more.

HUMIDITY The proportion of water vapor in air. Air with low humidity feels dry, while air with high humidity feels moist or damp.

HURRICANE A storm with average wind speeds of 74 mph (119 km/h) or more.

HYDROELECTRIC DAM A structure built across a river that allows a controlled amount of water to flow through to spin the angled blades of turbines. These are connected to electricity generators.

ICE AGE When global temperatures are low enough for large areas of the planet to become permanently covered in ice, in the form of glaciers, ice sheets, and ice caps.

INFRARED A form of electromagnetic radiation with longer wavelengths than visible light. Infrared rays carry heat energy.

JET STREAM A narrow band of fast-moving air in the atmosphere, usually at a height of between 5 and 10 miles (8 and 16 km). Jet streams greatly influence weather and climate.

LATITUDE A location's distance north or south of the Equator, measured in degrees. The Equator is located at 0° latitude and the North and South Poles are 90° latitude. London, UK, is about 51.5° north, while Sydney, Australia, is about 34° south.

LOW PRESSURE When the air is less dense and lighter, or "thinner," and so presses down less.

METEOROLOGIST A scientist who studies the atmosphere, including features such as its gases, temperature, air pressure, and wind, in order to make predictions about the weather and climate.

METEOROLOGY The scientific study of the atmosphere for the purpose of making predictions about the weather and climate.

METEOSAT Stands for "meteorological satellite"—a specialized space satellite that measures features of Earth's weather and climate, to assist with weather forecasting and to monitor global warming and climate change.

MICROCLIMATE The weather or climate in a small area, such as in a cave, among trees in a woodland, or on a small island.

MONSOON A period of heavy rain, usually for several weeks, which occurs around the same time, or in the same season, every year.

PHOTOVOLTAIC Capable of producing electricity from light energy.

PRECIPITATION Any form of water falling from Earth's atmosphere onto its surface, such as rain, hail, sleet, or snow.

RADAR Radio Detection And Ranging— a system for measuring the location, distance, direction, and speed of objects by using devices that send out radio waves and then detect the reflections—or echoes —that bounce back from the objects.

RAIN SHADOW An area of land that lies behind a mountain and gets very little or no rainfall. This is because the air loses most of its moisture as rain when it passes up and over the mountain.

RENEWABLE ENERGY Energy made from resources that nature will constantly replace, such as wind, water, and sunshine. Renewable energy is also known as clean energy.

SATELLITE An object traveling around, or orbiting, a larger object. The Moon is a natural satellite of Earth. The term is also applied to artificial, or man-made, objects such as spacecraft that orbit Earth.

SOLAR PANEL A wide, flat device that turns the Sun's energy into electrical energy. These panels are usually photovoltaic, which means that they change sunlight into electricity.

SOLAR THERMAL ENERGY The Sun's heat or warmth, in the form of invisible rays known as infrared radiation.

TEMPERATE A temperate climate is generally moderate all year round and does not experience wide extremes of temperature.

TORNADO An extremely fast-spinning column, or funnel, of air between Earth's surface and a storm cloud (cumulonimbus). Also called twisters or whirlwinds, they are usually less than 2 miles (3 km) across and usually form on land.

TRADE WINDS Generally regular, predictable, large-scale winds that blow from east to west, just to the north and south of the Equator.

TRANSPIRATION How water moves through a plant, entering through the roots and eventually exiting through holes in the leaves, evaporating into the air as water vapor.

TROPICAL The region between the Tropics of Capricorn and Cancer, on either side of the Equator, where temperatures are generally high all year round.

TROPICAL CYCLONE A very large-scale zone of atmospheric low pressure that forms in the Tropics, usually causing huge storms. These storms are known as hurricanes over the North Atlantic Ocean and northeast Pacific, typhoons over the western Pacific Ocean, and cyclones over the South Pacific and Indian Ocean.

TURBINE A machine with angled blades (like an electric fan) that spin on a shaft. The blades are turned by various forms of motion energy—for example flowing water or wind, to generate electricity.

TYPHOON The name given to a huge storm that forms in the western Pacific Ocean, caused by a tropical cyclone. Typhoons are known by different names in different parts of the Tropics.

ULTRAVIOLET A form of electromagnetic radiation, like infrared and visible light, but with much shorter wavelengths than both. It is present in sunlight. High exposure can be dangerous to human skin.

WEATHER The overall conditions created by the clouds, wind, temperature, humidity, and similar atmospheric features in a particular place over a period of days or weeks.

WEATHER FRONT The boundary, or transition, between two different large-scale masses of air, usually between low and high air pressure.

INDEX